Fiber Optics in
Communications
Systems

ELECTRO-OPTICS SERIES

Series Editor: Dr. Herbert Elion

Managing Director, Electro-Optics
Arthur D. Little, Inc.
Cambridge, Massachusetts

Other volumes in preparation

Fiber Optics in Communications Systems

GLENN R. ELION
Sumitomo Electric Industries, Ltd.
Yokohama, Japan

HERBERT A. ELION
Managing Director, Electro-Optics
Arthur D. Little, Inc.
Cambridge, Massachusetts

MARCEL DEKKER, INC.   New York and Basel

Library of Congress Cataloging in Publication Data

Elion, Glenn, [Date]
    Fiber optics in communications systems.

    (Electro-optics series ; v. 2)
    Includes index.
    1. Optical communications. 2. Fiber optics.
I. Elion, Herbert A., joint author. II. Title.
III. Series.
TK5103.59.E43   621.38'0414   78-2294
ISBN 0-8247-6742-X

MARCEL DEKKER, INC.

270 Madison Avenue, New York, New York 10016

Current printing (last digit):
10 9 8 7 6 5 4 3 2

PRINTED IN THE UNITED STATES OF AMERICA

D
621.3804'14
ELI

PREFACE

In the twentieth century, man has witnessed a great explosion in the
realm of electronic communications, especially with telephones, radio,
television, computers, and satellite communications. Each new method
of communication has brought new technology, new components, and cre-
ative manufacturing methods. Significant steps have been made in many
areas of the electronics industry to reduce the size and cost of many
items and components, and simultaneously increase quality, performance,
and service. Just when some new level has been achieved on the market
that seems to be an ultimate in size, cost, and performance optimization, something newer comes along.

In the twenty-first century, electronic communications and components will be based heavily on electro-optics. At that time it is
expected that all new telephones, computers, and television systems
will be linked by fiber optic cables (in conjunction with space satel-
lites) using laser light operating with integrated optical circuits.
Computer plotters and recording devices of today will largely be replaced by the instantaneous fiber optic plotters and display systems
of tomorrow. Industrial automation and control will use many fiber
optic systems bringing dramatic changes in required equipment standards, especially in the explosion-prone petrochemical industries.
By 1978, some of these seeming twenty-first century predictions will
already be physical reality or on their way into production.

This book discusses in detail fiber optic communications systems
at the present state-of-the-art. The fiber optic systems market is
the largest growing in the electro-optics industry. The major fiber
optic applications include telephones, cable TV, industrial automations, and computers. The size of the industry in 1978 is in the

millions and by the 1980s a multi-billion dollar industry will have evolved.

This book describes major components including fibers, cables, emission sources, detectors, modulators, and repeaters, as well as total system designs. Economics and business opportunities in fiber optic communication systems are discussed using existing communications equipment and capabilities and costs for comparison.

A significant effort was made to collect pertinent and interesting references which are listed not only with the title of the paper, but all are broken down into appropriate categories. This will be particularly useful to those professionals in the field who wish to get more information on their specific work areas, and to the novice to get more background on areas of interest which may require further details for better comprehension.

Since the field is changing so rapidly, and the technology has been continually updated, the oldest references included in this book are from 1974. The most recent references are as close to the printing date as possible to provide the most up-to-date information and achievements in the field of fiber optic communications systems.

There have been many companies who have contributed to this work by providing literature and technical product information. Some of this information, particularly on fiber specifications, has been included in the book. A list of suppliers of various fiber optic and electro-optic products has been included in several chapters. It is not a complete international listing, but it gives a sufficient number of companies for each product line to provide a source of further detailed information. New compnaies and products are continually springing forth. By combining the given lists with the leads that can be found by following the current trade journals for new products and companies, a complete source listing of all available materials can be compiled.

Glenn R. Elion
Herbert A. Elion

CONTENTS

Contents

LIST OF FIGURES AND TABLES

Figures

Tables

Fiber Optics in
Communications
Systems

INTRODUCTION

There is an ever increasing demand for efficient communication systems caused by increasing costs to supply more new equipment and increased consumer demand for all information services especially in television, radio, telephones, and computers. Using the equipment and technology of the past decade, many companies and business enterprises are having economic problems with increased costs for metal cabling, energy, and for finding available space on existing transmission lines and available carrier frequencies. Various new forms of telecommunications are appearing which require large bandwidths which put a considerable load on existing lines such as conference television, video-telephones, and viewdata systems. The use of more ultra-short wave links is approaching its limits in some areas due to a shortage of bandwidth and mutual interference.

Most systems are limited by one important factor, the information bandwidth. For wire and coaxial lines, the frequency available for information transmission extends from 10^8 to 10^9 Hz. Using multiplexing and coding techniques, the information bandwidth has been increased for these systems to a seemingly high level of information transmission per channel per unit time. However, using modulated light transmission where frequencies are in the 10^{10} Hz range, an increase in several orders of magnitude in potential bandwidth is possible.

In addition to the step jump increase in information bandwidth available per channel, there are several other advantages of optical transmission lines compared to conventional systems.

1

1. One fiber optic filament, several mils in diameter,
   can replace a copper wire cable several inches in diameter. This savings in size and weight is especially
   important for underwater cables and in areas of overcrowded transmission wires.

2. Optical transmission is immune to any ambient electrical
   noise, ringing, echoes, or electromagnetic interference,
   and does not generate any of its own electrical noise.
   Fiber optic communication does not present any problems
   of crosstalk.

3. Optical cables are safe to use in explosive environments and eliminate the hazards of short-circuits in
   metal wires and cables. Optical systems can be made to
   have total electrical isolation.

4. Properly designed optical transmission lines and couplers are relatively immune to adverse temperature and
   moisture conditions and can be used for underwater
   cable. Under very adverse conditions, bare fibers of
   glass (silica) composition can withstand 1000°C where
   coax cable is limited to ∿300°C extremes.

5. The number of repeaters required for low attenuation
   cable is less than with conventional systems and for
   short hauls of less than 10 km, no repeaters will be
   necessary.

6. Fiber optic cables, today, cost about the same as pre-
   mium grade coaxial cable. As production volumes increase, these costs will drop to below one-half their
   present cost level, making fiber cable very advantageous in terms of initial cost per unit length.

7. Equipment used now for electronic cable that protects
   against grounding and voltage problems, can be eliminated when using fiber cables. (When repeaters are used,
   especially in long distance underwater cables where
   power is supplied within the cables for repeaters, care

still must be taken to avoid voltage problems in
fiber optic systems.)

8. Most low-loss fiber optic cables made today can be upgraded by changing LED light sources to injection
lasers as they become available. Upgrading can also be
achieved by improved modulation techniques and equipment without replacing the original cable.

✓9. The installation costs of fiber optic cables are lower
than metal cables since the shipping and handling costs
are about one-fourth that of current metal cables and
labor about one-half less.

(These advantages, along with information bandwidth increases, make optical communications very attractive for numerous
applications including telephones (loops, trunks, terminals, and
exchanges), computers (internal and external links), cable television (trunks, distribution centers, and closed-circuit), space
vehicles, avionics (military and commercial aircraft), ships (military and commercial craft), submarine cable and special tethers,
security and alarm systems, electronic instrumentation systems,
medical systems (patient monitoring and surgical procedures),
satellite ground stations, and industrial automation and process
controls (chemical, oil and nuclear plants). (Of all these applications, the telecommunications and computer industries will account for about three-quarters of the electro-optics communication
market in the early 1980's.)

The simple LED fiber optic communication system is shown in
Figure 1. Depending upon the fiber cable attenuation, the length
of the transmission cable, signal strength and other factors, repeaters may or may not be needed. The receiving terminal end may
be a simple end-device or a station to switch, split and couple
signals to other destinations such as in telephone and cable television systems.

Simple communication links, such as depicted in Figure 1,
are already in existence and functional operation in many places

FIGURE 1

BASIC LED FIBER OPTIC COMMUNICATIONS SYSTEM

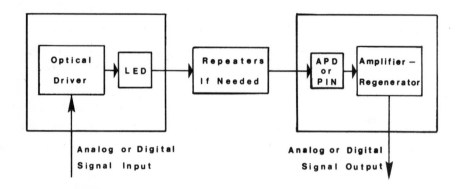

throughout the world, particularly in Europe, Japan, and the United States. Optical transmission, using lasers as the light sources, are not as yet as common. As modulation techniques and devices become more sophisticated and the lifetimes and costs of laser sources become more in line with existing LED's, the use of laser optical communications will eventually dominate the market due to the increased bandwidth potential.

The competition worldwide for recognition and dominance in electro-optic communication systems is rather fierce. During the coming few years, as more and more devices, systems, fibers, and components enter the public market, this industrial and technical competition will increase even further, causing prices of many items to decrease. Many small companies and peripheral business enterprises can be expected to be absorbed, merged, or put out of business by larger and more aggressive firms. More engineering schools will be adding special courses on electro-optic communication systems and devices, and more specialized educational programs for existing industrial personnel can be expected to emerge throughout the world.

The purpose of this book is to give the novice and the practicing industrial and research experts a text which presents the overall state-of-the-art, as well as specific design details for complete systems. In addition, this book presents practical as well as theoretical descriptions of available fibers, cables, couplers, connectors, splices, light sources, modulators, photodetectors and repeaters. Overall system design is depicted for various types of fiber optic communications applications. A listing of many companies producing or supplying fiber optic components and links is given in the Appendix as an additional source of specific material on existing equipment.

CHAPTER 1

FIBERS AND CABLES

I. Fibers

*The major characteristics to be considered in optical fiber
transmission lines are:

    1.   attenuation--and its variance with transmission input
wavelength, modal distribution and cable temperature.

    2.   distortion--and its variance with bandwidth, modal dis-
tribution, the amplitudes and wavelength of the input
light, length of the fiber and environmental temperature,

    3.   radiation--and its variance with bend radius and fiber
temperature.

    4.   physical parameters--including size, weight, total
volume, ease of installation, splicing and coupling.

    5.   environmental parameters--including resistance to water,
stress and chemical corrosion, mechanical stresses and
temperature.

A. Numerical Aperture

The numerical aperture of an optical fiber is defined as:
$NA = (N_{core}^2 - N_{clad}^2)^{1/2}$ where N_{core} is the refractive index of
the core and N_{clad} is the refractive index of the cladding.
Figure 2 depicts the light propagation from an open ended fiber
and the relationships of the indices of refraction. The optical
numerical aperture of a fiber depends upon the specific application.

FIGURE 2

LIGHT PROPAGATION FROM OPEN END FIBER

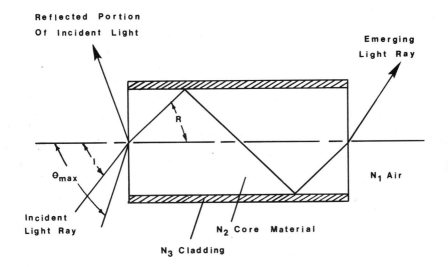

$$\text{Sin } \Theta_{max} = \text{ Numerical   Aperture}$$

$$N_1 = \text{ Refractive Index Of Air (1.0)}$$

$$N_1 \text{ Sin } (\Theta_{max}) = (N_2^2 - N_3^2)^{\frac{1}{2}}$$

The numerical aperture determines the coupling efficiency between the LED or laser light source and the optical fiber. Small values of NA tend to give low pulse dispersions; but, on the other hand, narrow confinement of the light to the fiber core enhances losses due to microbending. For long distance telecommunications fibers, an NA of about 0.2 is usually acceptable. For short distance applications, higher NA values are sometimes desirable.

The numerical aperture is determined by the type and concentration of dopants in the fiber. The oxides of germanium, boron, phosphorous, titanium, and aluminum are commonly used as dopants. For graded index fibers, NA values of 0.15 to 0.25 are common with plastic clad silica fibers having values about 0.20. For silica fibers doped with several percent fluorine, numerical aperture values can be as high as 0.25 with attenuations less than 4db/km. In general, fibers with large NA values have high losses but are easier and cheaper to manufacture since the fiber drawing process is simpler and the demand on chemical purity and imperfections is not as great as in lower loss fibers.

B. Index of Refraction

Choosing the absolute values of the index of refraction for core and cladding materials is not a simple matter, and depends in part upon the specific communications application. A trade-off exists in choosing these values in that increasing the difference in the refractive index of the core and cladding permits better source coupling; but, at the same time, increases the degree of intermodal dispersion (pulse spreading). Typical refractive index radial profiles are shown in Figure 3.

Using an approximate parabolic profile for the index of refraction minimizes modal propagation time delay differences (often called modal delay distortion) thus increasing bandwidth capabilities. A typical radial profile can be defined by:

FIGURE 3

REFRACTIVE INDEX RADIAL PROFILES

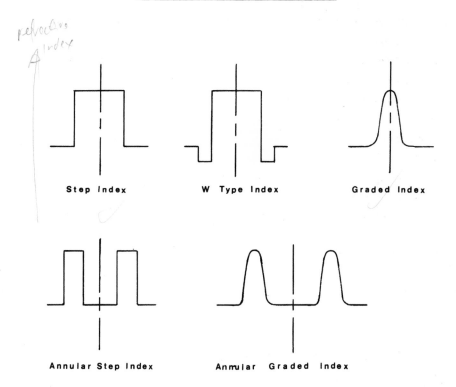

$$N_r = N_o \left\{ 1 - \frac{(NA)^2}{n_o} \left(\frac{r}{r_c} \right)^A \right\}^{\frac{1}{2}}$$

where n_o is the refractive index on the central axis, r_c is the radius of the fiber core and A is the profile parameter whose typical value is about 2.

Losses due to fiber bending decrease as the refractive index differential between the core and cladding increases and as the core radius decreases. However, as the index differential

increases, the waveguide bandwidth decreases, and as the core
radius decreases, the coupling efficiency decreases as well.

C. Attenuation

     Attenuation losses in fibers are caused by material ab-
sorption and scattering, waveguide scattering and radiation losses.
Material absorption is mostly caused by OH^- ions and transition
metal ions in the glass. Waveguide scattering is caused mainly
by geometric irregularities at the core-cladding interface. Carefully controlled fabrication procedures can keep waveguide scattering losses down to less than 1dB/Km. Radiation losses are
caused by the bending of fibers particularly at small radii of
curvature. Radiation losses can be particularly large when
cabling without plastic cushioning material surrounding the fibers.
These losses can be minimized in part by using high numerical apertures for the fibers which is difficult especially for graded index fibers without using some doping techniques to increase the
numerical aperture to 0.20 or beyond.

     To minimize fiber attenuations, certain material properties
are desirable in the waveguide. First, is to obtain a low concentration of impurities, particularly those which absorb in the
visible and near IR such as certain transition metals and especially OH^- groups. Second, is to match the dielectric properties
of all components, especially in multicomponent glasses, to minimize scattering from compositional fluctuations. Third, is to
achieve a relatively low glass transition temperature to minimize
density fluctuations caused by loss of volatile components and to
reduce the broadening of the UV absorption band.

     The effects of contamination by certain transition elements
in the fiber can be somewhat controlled by optimizing the redox
state of the glass to achieve a balance of the effects of various
metal ions in various oxidation states. For the design of specific
transmission systems, the fiber composition can be optimized for
     r wavelength to achieve the greatest transmission of

the light source along the fiber. When designing fibers for certain laser frequencies, it is possible to add buffering agents such
as As_2O_3 to the glass to control and stabilize the redox state of
the transition element contaminants. For Fe^{+2} and Co^{+3} ions, the
total waveguide concentration cannot exceed 1ppb to obtain lower
than 1dB/Km attenuation levels. The major impurity of concern in
silica based optical waveguides is the OH^- radical which has weak
absorption bands at 725, 825, and 875nm and a strong absorption
band at 950nm. Figure 4 shows the fiber attenuation versus wavelength for typical low loss fibers. This larger band contributes
to transmission losses at a rate of about 1dB/km/ppm. To minimize the OH^- absorption band in fibers, starting materials are
often heated and dried and used in very low humidity controlled
production areas. The limitation of using totally dry materials
and manufacturing environments is that defect absorption effects
can occur from oxygen defects initiated by the fiber drawing process.

       Other losses are caused by density fluctuations within the
fiber. This intrinsic type of scattering is the fundamental
limit to attenuation in waveguides. The magnitude of this scat-
tering loss is given by : $\alpha_s = \dfrac{8\pi^3}{3\lambda^2} \left(n^2 - 1 \right) KT\beta$

where T is the transition temperature at which the fluctuations are
frozen into the glass, β is the isothermal compressibility and
λ is the wavelength of the transmitted light. It is clear from
the $1/\lambda^4$ factor that this type of scattering loss decreases rapidly
with increasing wavelength. For fused silica with $T \sim 1700°K$, the
value of α_s is about 1-6 dB/km at 830nm. In multicomponent
glasses, the transition temperature can be lowered, thus reducing
the associated density fluctuation losses.

D. Pulse Dispersion

       The two types of dispersion in optical waveguides are
multimode group delay spreading and material dispersion, both

FIGURE 4

FIBER ATTENUATION VS. WAVELENGTH

FOR LOW LOSS FIBERS

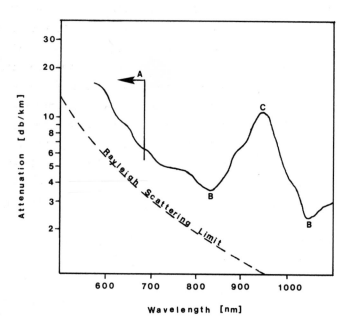

A — Drawing Induced Absorption Region
B — Low Loss Regions
C — OH⁻ Absorption Region

of which   limit the data rate capacity. Light, which enters a
fiber at an angle other than along the central axis, takes longer
to transverse the length of the fiber since it must undergo
numerous internal reflections per unit length of fiber with internal cores ranging from 10 to 100nm diameter. The difference
in the total distance travelled by central axis and off-axis rays
causes a time lag to occur which broadens the input pulse. This
pulse dispersion or pulse broadening effect limits the effective
bandpass and information capabilities of any fiber. By carefully
grading the change in the refractive index between the cladding and
core pulse broadening can be minimized. This decreases the effective path length differential between low and high order modes,
since larger angle rays travel faster through the lower refract-
tive index material portion of the waveguide reducing the time lag.

   To minimize pulse dispersion, the graded index profile in
fibers should be carefully tailored to the particular fiber material being used. For plastic clad high-silica fibers with step-
index profiles, pulse dispersions of about 20nsec/km can easily
be achieved for multimode fibers. Carefully prepared parabolically
graded index profiles can achieve dispersions as low as 0.2ns/km.
Graded index fibers made by the double crucible method using controlled metal-alkali in exchange at the draw nozzle have produced
fibers with pulse dispersions of about 1ns/km. Although the
parabolic index profile gives the smallest group delay distortion
compared to any other refractive index profile, it is very sensitive to physical deviations from the parabolic shape, making
manufacturing procedures very stringent.

   Group delay distortion in multimode fibers can be effectively reduced by making the differential in the index of refraction of the core and cladding less than 0.007. This, however,
makes the coupling of the input light source very difficult. In
single mode fibers with a refractive index differential of less
than 0.01, the signal dispersion becomes dominated by material dispersion instead of waveguide mode dispersion.

Pulse spreading for step index (SI) multimode fibers is given by: $\Delta_{SI} = n_{core} \Delta n L / c$, where Δn is the refractive index difference between the core and cladding and L is the fiber length. For graded index (GI) fibers, the pulse spread is given by: $\Delta_{GI} = An_{core} (\Delta n) L / 8c$ where c is the speed of light and A is a constant which related the different qualities of graded index profiles where the maximum value is 1.0. The ratio of pulse spreading for step-index versus graded index, fibers can theoretically (using A=1.0) achieve a value of $8/\Delta n$. In practice, this ratio ranges from 200 in single research grade fibers to 6 for production cables. Typical values of mode pulse spreading in graded index cables are 0.2-4.0 ns/km.

Material dispersion is caused by the nonlinear aspects of the refractive index with respect to transmission wavelength. The pulse spreading from material dispersion is given by: $\Delta_{mat} = \frac{\lambda(\Delta\lambda)}{c} \left(\frac{d^2 n_{core}}{d\lambda^2} \right)$ L, where $\Delta\lambda$ is the spectral width of the light source and λ is the average wavelength of the transmitted light. Typical material dispersion pulse spreading values for glass fibers using LED's in the 820nm window region are 3.0-3.5 ns/km ($\Delta\lambda = 380$Å) and for lasers operating in the 1040nm window are 0.2-0.3 ns/km ($\Delta\lambda = 20$Å).

For long fibers and cables (in excess of 300 meters), the pulse spreading per kilometer can be approximately extrapolated by relating the pulse spread to the square root of the fiber length. For short fibers, this relationship is linear (directly proportional) since there is little energy exchange between higher order modes over short transmission lengths. These relationships can be used to estimate the critical length or maximum length of a fiber waveguide which is determined by a pulse spread equal to the original pulse width for different bit rates (bps). At a bit rate of 10 Mbps for multimode fibers under step-index profiles, the critical length is approximately 25km using an LED and 50 km for a laser light source. For parabolic index profiles, the

critical lengths are 100km for an LED and 50,000km for a laser.
At a bit rate of 100 Mbps, these critical lengths are one-hundredth
the size for the same light source at 10 Mbps. From these approximations, it can quickly be seen how higher bit rates and
pulse spreading can severely limit the maximum length of an optical transmission communications cable.

In practice attenuation as well as pulse spreading limits
the length of transmission lines between repeaters. A comparison
of different transmission types, attenuations and resulting critical path lengths is shown in Table I.

E. Fiber Bending

The macro-bending of optical waveguides causes radiation
losses which become increasingly severe with decreasing bend radius.
The smallest permissible curvature radius is limited to the actual
fiber strength. Different fibers and cables have different
strain-to-failure properties depending upon construction methods
and materials and overall design. If a fiber is bent around a
surface of radius R (bend radius) the outermost edge of the fiber
cladding of radius r, will be strained relative to the fiber axis
by a certain percentage σ_s, where: $\sigma_s = \left(\dfrac{R+2r}{R+r} - 1\right) \times 100\%$. For
a bend radius of 1cm and a cladding radius of 70μm, $\sigma_s = 0.7\%$.
For fibers and cables to maintain longevity, the highest permissible differential bending strain should not exceed 0.2%, which
for the above fiber would mean a maximum bend radius of about 5cm.

When a manufacturer refers to a minimum bending radius, it
usually means the minimum radius that can be achieved before the
fiber breaks. This does not give an indication as to the expected
lifetime at slightly higher curvatures. For cable installations,
the widest permissible radii of curvature should be used to avoid
premature fiber failures.

Table I

Comparison of Different Transmission Types

Transmission Type	Attenuation (dB/km)	Critical Length(km)
optical	3 at 1um	10
coaxial	150 at 1GHz	0.5
microwave	150 at 10GHz	0.4

F. Fiber Materials

Advances in optical materials technology and fiber fabrication procedures have resulted in fiber waveguides with low transmission losses resulting from controlled fabrication techniques
and transition metal impurity levels less than several parts per
billion. The three major material systems used in the manufacture
of fibers are silica, glass, and plastic.

The silica system is basically made of silicon dioxide with
other metal oxides to establish a difference in the refractive
index between the core and cladding. Various dopants have been
used to increase the refractive index of silica including TiO_2,
Al_2O_3, GeO_2, and P_2O_5. Compound glasses used for cores and clad-
dings in optical fibers should have similar coefficients of thermal expansion, similar viscosities at the drawing temperature, low
material dispersion (high purity), low Rayleigh scattering (low
compositional and density fluctuations), low melting temperatures
and long term chemical stability. The cladding is generally of

lower refractive index and has a higher coefficient of thermal expansion than the core. Many dopants and materials used in waveguide production facilities are highly toxic and special precautions must be taken to handle and apply materials safely.

Chemical purity in silica systems is more difficult to maintain at high levels than glass systems since silica is insoluble in water and thus cannot employ the common purification processes such as in exchange and recrystallization techniques. Sodium aluminum silicate materials have a lower intrinsic scattering and can be manufactured more easily than pure silica. Table II lists a few material systems and the associated fiber properties.

G. Fiber Fabrication

A variety of fabrication techniques are available for optical fibers depending upon the core and cladding materials to be used and the desired refractive index profiles and attenuation limits. The major processes used are: double crucible (DC), stratified melt (SM) shown in Figure 5, chemical vapor deposition (CVD), rod-in-tube and direct fiber drawing.

All processes require ultra-purified starting materials. Some processes are batch type operations which first make preforms and subsequently make fibers by various drawing methods. A few processes can be designed to be continuous given sufficient supplies of raw materials, manpower and furnaces capable of maintaining constant temperatures for long periods of time. Such continuous processes are advantageous economically in terms of producing a large volume of long length fibers.

High-silica fibers are usually prepared by a gas phase reaction either by flame hydrolysis or by plasma induced chemical vapor deposition where doped layers of SiO_2 are deposited on the outside of a mandrel or on the inside of a fused silica tube. The prepared preform is then subsequently reduced to a rod and drawn out into the final fiber form and coated with an appropriate material. Various drawing methods can be used. Figure 6 shows

Table II

Materials Systems and Fiber Properties

Material System	Atten. (dB/km)	Fabrication Method	NA	Fiber Core Diam. (um)
SiO_2 - P_2O_5	2	CVD	0.18	50
silica-GeO_2	4	CVD	0.14	20
soda-boro- silicate	15	double crucible	0.15	20
alkali-lead silicate	30	cladded rod	0.45	50
soda-lime silicate	45	cladded rod	0.20	30

the basic laser fiber drawing process. Multicomponent fibers are prepared by mixing the core and cladding glasses separately and then drawing the rods directly from the melt. The rods are then subsequently drawn to fibers.

The chemical vapor-deposition fabrication method can be done by applying dopants externally to a silica mandrel by flame reactions of gaseous compounds or internally to a silica tube, as shown in Figure 7. The amorphous evenly coated mandrels or tubes are converted into a clear glass preform in a carefully controlled high temperature furnace and the preforms are then subsequently drawn into fibers. The preparation of preforms using gas phase reactions can be carried out with very high purity levels using the best available quality materials. Using carefully

FIGURE 5

STRATIFIED MELT PROCESS

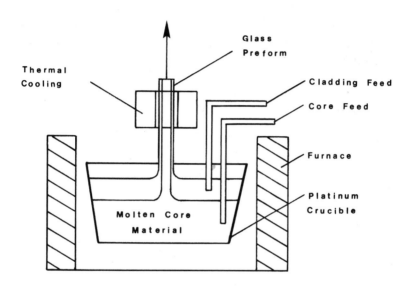

controlled preparation with controlled atmospheres fibers can
be made with losses approaching the Rayleigh scattering limit in
the neighborhood of 1dB/km at the 840nm window.

The conversion of preforms into fibers is interesting in
terms of the geometry of large rods drawn into thin fibers. The
physical drawing process is delicately controlled by interactive
process controls on both the furnace temperature and drawing
speed. A preform with a diameter (d_p) and length (L_p) can be
drawn into fiber of diameter (d_f) and length (L_f) as follows:
$d_p^2 L_p = d_f^2 L_f$. For a preform of 1.6m length and 30mm diameter to
be drawn to a fiber with a length of 10km, the maximum theoretical
outside fiber diameter would be 120um. In actuality, some material

FIGURE 6

LASER FIBER DRAWING PROCESS

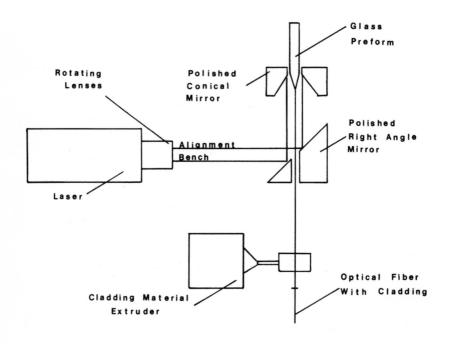

FIGURE 7

CHEMICAL VAPOR DEPOSITION PROCESS

FOR FIBER PREFORM MANUFACTURING

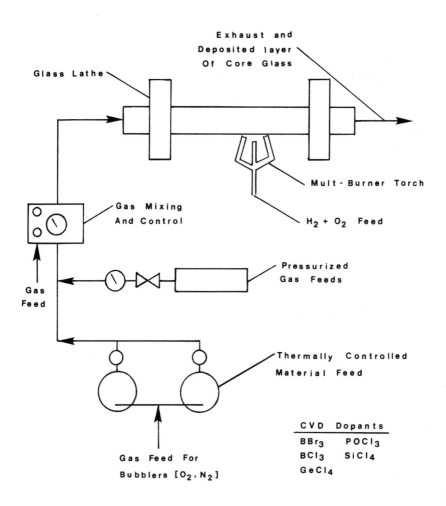

is lost at the beginning and at the end of the fiber drawing process such that a preform gives less than a 90% volume yield going from rod to fiber.

The double crucible (DC) method uses purified glasses in separate crucibles in a controlled atmosphere furnace, as shown in Figure 8. In some facilities about 1kg of glass is kept at the controlled melt temperature in each crucible, maintained by continuous feeding of additional glass rod material. The attenuations of some fibers have been field optimized by controlling the oxygen in the furnace atmosphere to balance the relative oxidation states of iron, copper, cobalt, and other transition metal ions. Some technical problems that occur with the double crucible method are bubbles forming in the melts and variations in the draw nozzle conditions causing minor fiber diameter fluctuations. When properly executed, the DC method can achieve fibers with graded refractive index profiles of 100-125μm diameter with ±1% geometric variations over the length of the fiber at production drawing rates of 1-10km/hr. The double crucible method has two advantages over the chemical vapor deposition method in that fiber production is continuous and that large values of the refractive index differential between core and cladding can be easily achieved. The DC method is not the panacea of fiber fabrication, however, in that the refractive index profile cannot be controlled as precisely as in other methods such as the chemical vapor deposition method and minor geometric variations at the cladding/core interface can initiate scattering losses. Polymer clad high-silica fibers have been prepared with losses of about 4dB/km and multicomponent glasses have been made with attenuations less than 9dB/km. Utilization of the double crucible method in making graded index borosilicate fibers has produced fibers with attenuations of about 5dB/km and step index fibers of about 10dB/km.

The stratified melt (SM) fabrication method works by floating the cladding material on the core material in a polished crucible and then drawing a preform from the doubled-layer melt. Since one

FIGURE 8

DOUBLE CRUCIBLE FIBER DRAWING PROCESS

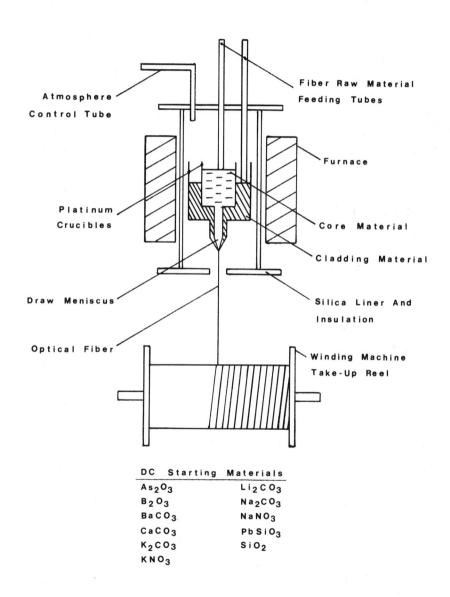

DC Starting Materials	
As_2O_3	Li_2CO_3
B_2O_3	Na_2CO_3
$BaCO_3$	$NaNO_3$
$CaCO_3$	$PbSiO_3$
K_2CO_3	SiO_2
KNO_3	

glass must essentially float on the other, the stratified melt
method is used mostly for step-index fibers. By continually
feeding the crucible with core and cladding raw materials, it is
possible to make this a continuous process by drawing the fiber
directly from the melt instead of using the preform intermediate
step. This method can produce fibers of a relatively large numerical aperture.

H. Coatings

In the coating of polymer materials onto optical fibers,
the coating must not induce any microbends into the fiber along
the entire length of the fiber and the particular plastics chosen
for the coating must protect the fiber physically and chemically
from the outside environment as well as mechanically from applied
stresses.

Most optical fibers which contain silica have a small relative elongation compared to polymeric coatings. Thus, the larger
the Young's modulus of the plastic coating, the thinner the applied
coating should be on the waveguide. The maximum coating thickness
is mostly determined by the coefficient of thermal expansion of the
applied polymer and the technical capabilities to apply uniform
thickness throughout the length of the fiber.

The total mechanical strength of the fiber is mainly determined by the condition of the surface. Initial manufacturing induced cracks propagate in the presence of applied mechanical stress.
Corrosion by surface moisture and certain acids accelerates this
process. The purpose of coating fibers is to protect the surface
from chemical attacks and mechanical damage or to at least minimize
these effects. Fibers of the best mechanical integrity are obtained by coating the fiber during the drawing process with a thin
plastic film of carefully controlled diameter to avoid intro-
ducing fiber microbending. When properly applied, coatings cause
no additional attenuation and increase mechanical strength and
stability compared to the naked fiber.

Numerous polymeric materials can be employed as coatings for optical waveguides applied in a variety of methods including dip-coating in solutions, extrusion, spray coating, and electrostatic coating. Typical plastics include teflon, FEP, PFA, and polyurethane. Plastic coated silica fibers often made with a perfluorinated ethylene propylene (FEP) coating on silica, are relatively inexpensive to manufacture, attain reasonably high NA values (0.20-0.25) and have attenuations ranging from 10-50dB/km. All plastic fibers (PCP) have very high attenuations no matter what coating material is used, but have high NA values (0.4-0.6). One advantage of PCP fibers and cables is total immunity to any induced water effects such as the stress-corrosion problems associated with fibers containing silica. Special plastic coatings can be used to protect the fiber from unique environmental and corrosion problems such as running fibers in locations containing acids or alkalis.

The strength of a given coated optical fiber depends upon the combination choice of plastic coating and the type of glass used. Given the same applied stress to a set of optical fibers, the effectiveness of several coatings ranked in increasing strength protection are: teflon, acrylics, silicones, and epoxies. The glasses ranked in increasing time before failure to a given applied stress are soda-lime, borosilicate, lithium aluminosilicate, and fused silica. The difference in the type of coating used can make a difference of about a factor of 10 in the failure time rating of the internal fiber-coating combination. The differences in glasses, however, are very large compared to the coating type. For the four glasses listed, there is an approximate order of magnitude increase between each one in the time before failure ratings to high applied stresses.

In the design of fibers and cables, the lifetime expectancy of the internal materials is only one consideration among many in making a final decision as to the composition and fabrication of a specific communications system.

I. Geometric Variation

The important geometric parameters to be considered during
the drawing process and fiber formation are to maintain constant
core and cladding diameters and exact coincidence of the circle
centers along the entire length of the fiber. Fluctuations of the
core diameter cause mode coupling. Changes in the outer diameter
of the fiber cause mechanical stresses and subsequent microbending
during the cabling process. The geometric variations using the
double crucible method are mostly determined by the physical drawing
process where other methods using intermediate preforms have added
potential irregularities.

Variations in fiber diameter and cross-sectional area can be
caused by minute changes in drawing tension, preform diameter, draw
speed, material composition, coating and curing, and take-up
spooling. Fiber drawing using induction furnaces can achieve rates
that exceed 10 meters/second while achieving geometric variations
of less than 0.4% using carefully controlled drawing with controlled atmospheres in the heat furnace.

J. Plastic Fibers

Optical plastic fibers have higher attenuations than glass
and silica material systems. They are commonly used for short distance computer applications with information capabilities of about
6 Mbps over distances of 50-200 meters. One distinct mechanical
advantage of plastic fibers is simpler and more reliable connections
and coupling since the plastic cladding can be gripped directly and
crimped with various connector ferrules and rings without cracking
or splitting. Plastic fiber couplings have attenuation losses
which are minute compared to the total fiber attenuation. Polymer
waveguides can be easily manufactured to have high numerical apertures.

Most plastic fibers and cables have operational temperature
ranges less than those of wire and coax cables. But, for some low
temperature applications, plastic fibers have a slight mechanical

strength advantage to dynamically applied stresses especially compared to some low-loss glass and silica fibers. For high temperatures or long distance applications, plastic fibers and cables cannot be used.

A listing of various types of commercially available fibers is given in Table III, with various fiber and cable parameters. This listing is not meant to be a complete listing even for the individual companies but rather is meant to show the diversity of fiber and cables and range of parameters currently available. Table IV gives a more complete listing of the companies who manufacture or supply fibers and cables.

K. Fiber Strength

The integrity of glass and silica optical fibers is highly susceptible to flaws in its structure since there is minimal yield under tension. Any defect whose chemical or physical structure leads to large local stresses is a weak spot or is a potential point of failure. In general, flaws in optical fiber can be characterized by: $\sigma_{max} = 2\sigma_{as} \left(\dfrac{c}{r_c} \right)^{1/2}$ where σ_{max} is the maximum local stress, σ_{as} is the applied stress, c is the equivalent flaw depth and r_c is the effective crack or flaw radius. When the local stress reaches a critical value determined by the weakest flaw in the fiber length, fracture occurs. The longer the fiber length, the weaker the flaws that determine any given failure rate since flaws are not evenly distributed throughout the entire length of the fiber. Since optical fibers can be up to 10km in length, it is very important to minimize the generation of flaws during manufacture, especially random weak points caused by imprecise controlling of all of the fiber drawing and coating parameters.

Optical waveguide communication systems, in many cases, are expected to last many years before need of replacement. Especially for cables which are difficult to install, such as underwater cables, it is important to know the expected lifetime-in-use of the

Table III

Parameters of Some Typical Commercial Optical Fibers and Cables

Manufacturer And Fiber ID Number	Fiber Material	Fiber Diam. (um)	Cable Diam. (mm)	# Of Fibers	Fiber Atten. (db/km)	Tensile Strength (kg)	Bend Radius (cm)	Index Profile	Max. Length (km)	NA
Bell North. BNR-7-1-A	Silica	100	--	1	10	30	3.0	Graded	0.5	.22
Bell North. BNR-7-2-A	Silica	100	--	1	10	30	3.0	Step	0.5	.20
Bell North. BNR-7-2-B	Silica	100	--	2	10	90	5.0	Graded	0.5	.22
Belling & Lee L2256-UPH	Glass	100	2.0	37	100	50	2.0	Step	0.02	--
BICC 1108, 1109	Silica	--	2.7	1-6	7	50	20.0	Step	1.0	.18
Corning 1028	Silica	125	--	1	6	10	2.5	Step	10.0	.18
Corning 1152-3	Silica	125	--	1	10	10	2.5	Graded	10.0	.20
Corning 1156-7	Silica	125	5.0	7	6	10	2.5	Graded	10.0	.16
Corning Corguide	Silica	125	5.0	7	20	200	15.0	Step	1.0	.18
Dupont PFX-S120R	Silica	600	2.4	1	50	80	0.3	Step	1.0	--
Dupont PFX-P740	Plastic	375	--	7	470	11	0.3	Step	1.0	--
Dupont PFX-P140R	Plastic	400	--	1	470	40	0.1	Step	1.0	--
Dupont PFX-P240R	Plastic	400	--	2	470	90	0.1	Step	1.0	--
Fiber Optic Q1-1-10	Silica	125	--	1	20	10	2.0	Step	0.01	.25
Fiber Optic Q1-7-5	Silica	125	--	1-7	50	30	2.0	Step	0.01	.25
Furukawa Elect. #22	Silica	125	--	1	6	5	2.0	Step	1.0	--
Furukawa Elect. #122	Silica	125	8.0	6	6	30	15.0	Step	1.0	--
Galileo 2000	Galite	70	4.2	210	400	26	2.0	Step	1.0	.66
Galileo 3000	Galite	110	4.2	1-19	100	26	2.0	Step	1.0	.48
Galileo 4000	Galite	200	4.2	1-19	30	26	2.0	Step	1.0	.40
Galileo 5000	Galite	125	4.2	1-19	20	26	2.0	Step	1.0	.20
General Cable AT	Silica	125	26.0	30	5	400	30.0	Graded	1.0	.21
ITT PS-05-10	Silica	500	--	1	10	150	0.3	Step	1.5	.30
ITT GG-02-5	Silica	125	--	1	5	150	0.5	Graded	1.5	.25

ITT S1	Glass	500	2.5	1	10	45	2.5	Step	1.0	.28
ITT LD	Glass	500	2.5	7-19	10	100	2.5	Step	1.0	.28
Klinger S-800	Silica	50	23.0	440	800	50	0.6	Step	2.0	--
Klinger A	Plastic	1015	---	1	1250	10	2.0	Step	2.0	--
Optics Research YDA	Silica	200	2.4	1	60	10	0.3	Step	0.02	--
Optics Research IDA	Glass	75	4.1	500	1000	50	1.5	Step	0.02	--
Optics Research SDB	Plastic	250	---		470	90	0.3	Step	0.02	--
Poly-Optics 1010	Acrylic	200	---	1	1100	150	0.4	Step	3.0	.51
Poly-Optics 1010C	Acrylic	500	---	7	1100	150	2.5	Step	3.0	.51
Quartz & Sil. 200A	Silica	400	3.0	1	5	10	0.3	Graded	10.0	.22
Quartz & Sil. CGI-60	Silica	150	3.0	1	8	30	5.0	Step	1.0	.25
Quartz & Sil. OFB-100	Glass	85	16.0	19-80	100	50	5.0	Step	1.0	.49
Rank Prec. Fibrox	Glass	50	2.3	1	700	10	3.0	Step	0.2	--
Rank Prec. Fibroflex	Glass	75	2.0	96	350	180	2.0	Step	0.2	--
Schott LK4-70	Silica	70	6.5	-	800	60	2.7	Step	0.2	.55
Siecor SIL-K4-15-10K	Silica	125	6.5	4	10	50	5.0	Graded	0.2	.20
Siecor SIL-K4-15-30K	Silica	140	6.5	4	30	50	5.0	Step	0.2	.20
Sumitomo Sumiguide PT	Silica	150	18.0	1-6	7	100	25.0	Graded	4.0	.25
Sumitomo Sumiguide CT	Glass	60	18.0	1-6	10	100	25.0	Graded	4.0	.20
Thomson Brandt ET1017	Plastic	85	9.4	19	100	55	15.0	Step	0.1	--
Thomson Brandt ET1018	Glass	85	9.4	37	100	55	15.0	Step	0.1	--
Times Fiber GP10-SAIO	Glass	125	8.5	10	10	50	10.0	Step	1.0	--
Times Fiber GP6-SAIO	Glass	125	6.4	6	10	50	6.4	Step	1.0	--
Times Fiber SAI-55-90	Glass	125	---	1	7	50	0.4	Step	1.0	.16
Valtec HHV-RT03-45	Glass	75	5.3	180	400	50	1.0	Step	1.0	--
Valtec LH-PC05-07	Silica	200	3.2	7	40	10	1.0	Step	1.0	--
Valtec MHV-MG05-01	Glass	125	4.0	1	10	10	1.0	Graded	2.0	.20

Source: Manufacturers' product sales literature, 1977.
N.B. For further and latest data, contact manufacturers.

Table IV

Some Manufacturers and Suppliers
of Optical Fibers and Cables

AEG-Telefunken	Hitachi Cable
American Optical	ITT
Anaconda Wire & Cable	Keystone Optical Fibers
Asahi Chemical	Klinger
Bell-Northern	Mitsubishi Rayon
Belling & Lee	NEC
BICC	Optics Research
Cables de Lyon	Pilkington Brothers
Canada Wire And Cable	Pirelli Industries
Corning Glass	Poly-Optics
Dainichi Nippon Cables	Quartz Products
Dolan-Jenner Industries	Quartz & Silice
Dupont	Rank Precision Industries
Dyonics	Schott
Fiber Optic Cable	Siecor
Fort	Siemens
Fujikura Cable Works	Sumitomo
Furukawa Electric	Thomson Brandt
Galileo Electro-Optics	Times Fiber Communications
General Cable	Valtec

cable and the individual fibers. In the presence of moisture and stress, flaws propagate and weaken the fiber. This makes testing difficult especially in terms of extrapolating expected time of failure for applied stresses over long periods of time.

Extended laboratory testing has concentrated on high stress loadings for periods of time extending to about three months. Practical low stress testing for greater periods of time will be determined empirically by field testing of operational installed systems. It is possible that the longer term stress corrosion factors will present themselves as problems for extended life of fibers and cables. Unfortunately, due to the wide variety of fiber and cable manufacturing procedures and products, this aspect of electro-optic communication systems may not be clearly defined for many years to come.

There are numerous means of determining some of the parameters used for lifetime and failure probability predictions, some of which gives a diversity of predicted values. Differences arise in testing methods due to non-uniformities in sample lengths, testing environments, and the means for gripping the fibers. Some significant differences also occur in using static and dynamic stress testing. The type and environment of the fiber strength testing should come as close as possible to simulating the actual intended field environment and applicaton instead of attempting to invent static testing methods which may be inappropriate for real cable situations.

Available fibers and cables have a very wide range of failure stress ratings ranging from a few thousand psi to more than 100 thousand psi. Failure is usually caused by the presence of surface defects and flaws ranging in size from .01μm to 10μm. To relate failure rates to applied stresses with various lengths of fiber over a period of time, Weibul statistics are commonly used. These statistics define a probability of failure (F) from 0 to 1, for fibers of length (L), under a stress (σ), over a period of time (t). $F = 1 - \exp\left\{ -\left(\dfrac{\sigma}{\sigma_o}\right)^m \left(\dfrac{t}{t_o}\right)^b \left(\dfrac{L}{L_o}\right) \right\}$, where

m is a distribution shape parameter and b is a parameter related
to susceptibility to stress corrosion. Once the values of m and b
have been established empirically over a variety of stresses, times
and fiber lengths, extrapolation can be made for any combination
of these factors. This is particularly valuable in extrapolating
to long time periods which could not be done in short-term laboratory experiments. Figure 9 shows typical plots for fiber failure
versus time for single silica fibers.

It must be kept in mind that when extrapolating over wide
ranges of time (years) or length (1-10km) that m and b are not
true constants, but rather are functions of time, length, and
strength among other factors. Approximations, however, are made
assuming m and b to be constant by: $\sigma_1^m \, t_1^b \, L_1 = \sigma_f^m \, \sigma_f^b \, L_f$ where
the subscripts 1 and f refer to laboratory and field values.

The applied stress to optical fibers also effects the
distortion losses as a function of fiber numerical aperture. For
most fibers the higher the numerical aperture the lesser the effects of stress induced distortion losses. The magnitude of stress
induced distortion losses also depends upon the fiber quality and
cabling manufacturing methods used to protect the inner fiber core.

L. Fiber Bundles

Cable fiber bundles are generally considered to be divided
into two types. One type contains a small number (5-50) of low
loss, low numerical aperture (100-1000) of high loss fibers with
a relatively large numerical aperture (.40-.70). Fiber bundles
are used to avoid the breakage problems of single fiber links which
can interrupt information transmission. By bundling numerous
fibers together with all sharing the same light source even if a
few fibers break, many will still be left to carry the signal. The
advantages of this bundling technique are added cable strength

FIGURE 9

TYPICAL FAILURE VS. TIME PLOTS

FOR SINGLE SILICA FIBERS

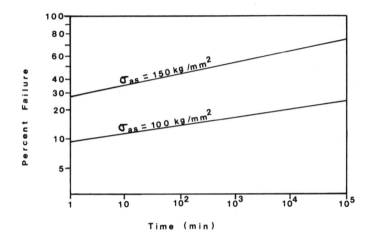

Fiber Length= 2.5 m., Diameter = 125 μm

and added cross-sectional area for the coupling of the bundle
to the light source.

In general, fiber bundles have high attenuations. High
loss fibers (200-2000dB/km) are inexpensive, reliable, have
large area light sources and may be considered to have a high
collection efficiency, and they are generally totally non-metallic
in composition. They are usually limited to short distance applications such as process control links, traffic control signs, and
certain computer applications.

Fiber bundles can be prepared by using an array of small
diameter glass preforms in the drawing furnace. Using this

simple technique, bundles ranging up to 1000 fibers can be made
in a single manufacturing step. The inherent technical problems
with this method are maintaining constant melting temperatures
for all rods and constant draw on all fibers. As a result, the
attenuations of fiber bundles presently range from 50 to 2000
dB/Km. As techniques improve, it should be possible to achieve
multifiber bundles with attenuations in the range of 20-30 dB/Km.

II. Cables

The requirements for fiber optic cables are as follows:

1. They can be handled in the same way as most ordinary
    communications wire and coax cables.
2. They have mechanical and electrical properties com-
    patible for their specific use and surrounding condi-
    tions, including delivered power to repeaters and
    resistance to applied static and dynamic stresses.
3. They can be permanently spliced and connected in the
    field with reasonable ease within a reasonable period
    of time.
4. They are economically competitive with existing wire
    and coax communications cables.
5. They have low optical attenuation approximately equi-
    valent to the individual fibers.
6. The individual fibers within the cables must be dis-
    tinguishable for identification.

The three important types of stress that must be considered
when designing an optical cable are tension, torsion, and flexure.
The wrapping of fibers around a central core such as a metal
strength member can induce tension as well as torsion stresses.
To increase the maximum tensile strength ratings of cables,
strength members are incorporated within the outer cable sheath
in various symmetrical of asymmetrical formats, as shown in
Figure 10.

Strength members added to the cable must have a high tensile

FIGURE 10

VARIOUS CABLE CONFIGURATIONS

PC— Plastic Cushioning
PF— Plastic Fillers
PS — Plastic Sheath
PT — Plastic Tape

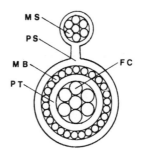

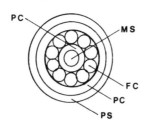

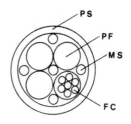

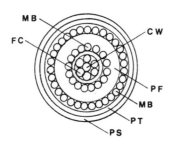

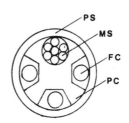

CW— Conducting Wire
FC — Fiber Cables
MB — Metal Braiding
MS — Metal Strength Member

strength, a high Young's modulus and high flexibility. There are
many materials with high tensile strength, but in order to achieve
flexibility as well, the material should be malleable enough to
be woven into multi-filament strands such as steel, copper, and
some specialized plastics such as Kevlar. Figure 11 depicts the
stress versus strain plots for various strength members materials.
In some cases, such as potentially explosive atmospheres or where
electro-magnetic interferences are strong, metal strength members cannot be used and must be substituted with members such as
a polyurethane coated Kevlar multi-strand filament.

The three basic designs used for cables are the ribbon,
strand, and internal cavity types. There is a wide choice of
plastic coatings available for the fiber and a large variety of
strength members both plastic and metal that can be used for the
cable design. It is difficult to design cables for general usage
due to the large variations encountered in environment and bandwidth requirements.

The multiplicative cost factor of cabling N fibers is less
than one, meaning that N fibers can be enclosed within the outer
protective sheath of a cable cheaper than making N cables. Within a limited range, the cost of multiple fiber low-loss cables can
be estimated by taking the cost of cabling a few fibers N_o and
multiplying the cost by $2N/3N_o$. This rule of thumb for cabling
costs applies generally in the range of 5-50 fibers, where each
low-loss fiber is connected to a separate light source. The prices
of cables from various manufacturers ranges from $1-$10/m mostly
depending upon cable attenuation ratings and the number of individual fibers contained within the cable; and to a lesser degree,
by the dispersion ratings of the fiber waveguides. The choice of
using graded index or step index fibers within the cables is
basically determined by the desired bandwidth-length capacity of
the communications system. High capacity systems must employ
graded index cables.

The projected cost evaluations of fibers and cables indicate

FIGURE 11

STRESS VS. STRAIN FOR VARIOUS

STRENGTH MEMBER MATERIALS

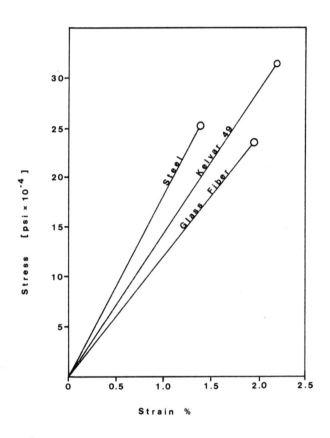

lower prices as large production facilities are put into operation
and as stiffer international competition for fiber communications
devices emerges. Prices below $.50/m for low loss cables should
be seen in 1979. Some initial concern that the dramatic rises in
energy costs would outweight mass production cost reduction has
proven unfounded in that utilities costs for most fabrication
methods accounts for only 4%-7% of the total manufacturing cost.

A. Stranding

The following are some of the basic requirements for the
stranding of fiber optic cables.

1. The manufacturing methods must be economical to main-
    tain the overall cost competitive nature of optical
    cables compared to coaxial cables.
2. The cable must be designed to allow reasonably quick
    and accurate jointing in the field.
3. Microbends can be induced into the fibers during the
    stranding process, so manufacturing controls must be
    stringent to prevent unnecessary losses.
4. The applied mechanical stress to the cable during
    manufacture and installation must be minimized.

The two basic types of stranding for fiber cables are
the layer and unit methods. In the unit method, coated fibers
are stranded into small groups and then the groups are stranded
together into larger groups to eventually make the agglomerate
bundle. In the layer method, the fibers are stranded in horizontal
or concentric layers to form the cable core. For both methods,
a cushioning material is placed between the layers or groups to
absorb applied stresses, especially those which are perpendicular
to the fiber axis.

B. Cabling Losses

The two mechanisms that increase attenuation when fibers
are made into cables are bending losses and microbending. The

bending losses are usually small compared to the individual fiber
attenuation in most flexible cables. Microbending losses are
caused by small asymmetrical forces acting on the fibers which have
a small diameter and can be distorted greatly by small forces.
Minor imperfections in all cabling surfaces causes some microbending losses in cabling processes. To minimize attenuation in
optical cables, the fibers can be loosely packed in open channels
made from soft flexible materials, or the fibers can be embedded
in a cushioning material that absorbs various applied stresses.
Cable losses due to bending can be greatly reduced by increasing
the cladding diameter and NA of the fibers and by decreasing the
core to cladding diameter ratio.

 Attenuation may be considered a length dependent loss as
opposed to one-time end losses which occur at the junction points
in the transmission system. These end losses consist mainly of
Fresnel reflections at both ends or joints of the cable. For very
short cables such as process control links and computer links, the
end losses are often a major source of transmission losses. Large
area light sources can be used with a number of fibers in parallel
for short links to keep transmission rates high with a high safety
factor and low BER. For long cables where attenuation losses dominate, the use of multiple fibers for the same channel is not practical. In long haul fiber optic cables which employ numerous
couplings, connections, splices and signal splitting, the "one-
time" losses can become rather significant and can sometimes be
greater than carefully designed low-loss cables. For such systems,
it is important to use cables with the lowest attenuation possible
to minimize the total system losses.

 Various mathematical models have been developed to predict
expected losses as a function of fiber physical parameters in
coating and cables. Unfortunately, most models only apply to
specific sets of fiber materials and conditions. Most models only
apply for step-index or for graded index fibers made of glass or
silica where expected losses are less than 50dB/km. High loss

cabling procedures or plastic fibers cannot have general descrip-
tive models due to the vast variety and magnitude of induced losses.

A typical model for low-loss fibers should include the
following variables: fiber radius, core radius, refractive index
differential, material modulus of fiber and coating, and a factor for induced cabling bumps or axial and radial geometric deviations. One such model is as follows:

$$\text{loss dB/km} = \frac{.9 R_c^4 B_n B_h^2}{(\Delta n)^3 R_f^6} \left[\frac{E_c}{E_f} \right]^{3/2}$$

where R_c is the core radius, R_f is the fiber radius, Δn is defined as the refractive index differential between core and cladding, B_n is the linear average number of occurrences of irregular
bumps of average height B_h. The material modulus of the core is
given by E_c, and the fiber by E_f. Some typical values for the
above parameters are as follows:

R_c = 50μm, R_f = 125μm, B_n = 200/km, B_h = 2μm

Δn = 0.12, E_c = 2x10^{5}psi, E_f = 8x10^{6}psi.

When substituted into the above model, the expected calculated
loss is 3.8dB/km. Note that this calculated loss is in addition
to expected material attenuations for a perfectly straight homo-
geneous fiber.

Distortion losses in cables are a function of the stress
actually experienced by the fiber inside the cable and the fiber
numerical aperture among other factors.

C. Underwater Cables

The design of underwater communications cables must take
into account numerous factors including cable diameter and weight,
the weight and volume of the available cable handling and storage
units, power transfer, bandwidth, submersible high capacity repeaters and modulators, environmental effects such as high pressure
and salt water and system lifetime expectancy and upgradeability.

A properly designed transoceanic optical telephone cable would be smaller, lighter, easier to install, cheaper and have a bandwidth capability orders of magnitude greater than existing coax cable systems. These facts are more than sufficient incentive to divert future efforts into employing practical fiber optic submarine communications cable. Such cables must be made to face the problems of contamination by water, axial strain and radial stresses, fiber microbending and cable and fiber macrobends.

Bare uncoated optical fibers containing silica weaken in the presence of water while under mechanical stress. Surface imperfections propagate and eventually cause fiber discontinuity. Glasses and silicas do not exhibit this behaviour in the presence of water (hydroxyl ions) when they are not under applied stresses. To minimize this stress corrosion, underwater cables should be designed to keep water from contacting the fibers, and should be deployed to maintain minimal axial stresses on the cable. Depending upon where the cable is deployed, its density and total auxiliary weights including repeaters, oceanic cable can be exposed to external water pressure exceeding 800kg/cm^2.

The material coatings commonly used for fibers to be employed in underwater cables are teflon or a polyester elastomer with an outer protective jacket of polyethylene. These plastics are impervious to water and to most common oceanic corrosion environments. For ocean cables, the lifetime of the cable must be optimized as best as possible since it is difficult and expensive to replace or repair cables. One means of extending fiber lifetime is to coat the optical waveguides as soon in the drawing process as possible when the fibers have been cooled below $350°C$ in a low humidity controlled environment. By doing this, the fiber surface is exposed to a minimal amount of water vapor and potential future stress corrosion. The thickness of the plastic coating may be slightly larger than those used for surface cables to add an extra margin of safety. To prevent any induced fiber microbends, the outer surface of the plastic coating must be smooth

and in concentric alignment with the internal fiber cladding and
core. The coating cannot be too thick or stiff otherwise it will
lose its ability to alleviate local concentrations along the fiber
surface.

The fiber bending problems with underwater cables presents
some interesting and unique situations. On surface cable applications, the cables can be lain on flat surfaces such as building
conduit or strung along underground pipes or telephone poles.
These situations allow the cable to maintain reasonably high curvature radii. However, in underwater cable deployment, there are
sometimes no guarantees as to what surfaces the cables will
eventually settle upon especially in deep water situations. Submarine cables may lay over rocks, vegetations, coral, or small
chasms and are exposed to being used by various ocean organisms
as an anchoring surface. All of these can put unexpected curves
into the cable. One way to eliminate radiation losses by this type
of bending is to use strengthening members within the cable to increase its overall inflexibility such that it will not limply
conform to every surface irregularity. The minimum bend radius
for underwater cables designed for longevity should not be allowed
to be below 6cm for fibers with cladding diameters less than
150µm.

Depending upon where the cable is to be placed, it can be
designed to be buoyant or to sink to the bottom. Those cables
which are designed to float above the ocean floor must take into
consideration underwater currents and the possibility of being
struck by marine animals or submersible craft. Heavy duty multi-
channel submarine cables using high-density plastic sheaths in-
evitably have sufficient weight to contact the ocean floor.
Lighter and smaller cables can be made with various filler materials to float at any desired level.

For cables that must transverse greater distances than
10 km repeaters must be used. For shorter links, such as some
island to island or land to island runs, no repeaters will be

needed. Where repeaters must be used, electrical conduit of some
sort must be included inside the sheath to power the repeater
units. Steel, copper, or aluminum wires could be used depending
upon the requirements for ultimate tensile strength, cable weight
and electrical resistance which for very long cable becomes a
significant factor.

In Table V a comparison is made of a typical coaxial telecommunications submarine cable system versus an optical fiber
underwater cable. The calculations are based on a 5000km cable
transoceanic run using existing bandwidth capabilities of coax
of optic fiber transmission lines. The number of channels for
the optical cable has been conservatively chosen at 600 with the
understanding that laser light sources instead of LED's and high
bandwidth capacity repeaters and modulators with multiplexing
could greatly increase this number for heavy duty multifiber
cables.

III. Measurements

The major parameters most commonly measured for the evaluation of fiber optic transmission system performance are as follows:

1. reflection coefficient--the ratio of the forward-
   to-reverse power along the transmission line at a
   given location in the line.
2. signal level--the peak and average carrier power
   contained within the waveguide at a given location
   in the line.
3. loss--the difference between transmitted and received
   power determined by fiber attenuation, various coupling
   and slicing losses and component losses.
4. pulse spread--the change in arrival time of various
   signal components at a given location in the line
   caused by material and modal dispersions.

Table V

Coaxial Vs. Optical Underwater Cable Systems

System Parameter	Coaxial Cable	Optical Cable	Ratio - Coax To Optical
cable diameter (cm)	3.2	1.6	2.0
cable weight (kg/km)	1250	625	2.0
total cable weight (kg)	6×10^6	3×10^6	2.0
total cable volume (m³)	4021	1005	4.0
cost of cable ($M)	35	25	1.4
number of repeaters	150	500	0.3
cost of each repeater ($K)	60	10	6.0
cost of all repeaters ($M)	9	5	1.8
total cable cost ($M)	44	30	1.5
number of channels	300	600	0.5
cost per channel ($K)	147	50	2.9

---

5. signal-to-noise ratio--the ratio of the power of
   the desired transmitted signal to all undesired
   power in the channel such as thermal noise.

6. impulse noise--defined as sudden extraneous signals
   in excess of 12dB of the rms noise level caused by
   electrical problems with the light source, detector,
   modulator, or switching components.

7. temperature resistance--the ability of the transmission
   system to withstand temperature fluctuations usually
   rated by maximum and minimum operating temperature
   limits and variations of transmission attenuation
   as a function of temperature.

8. corrosion resistance--the resistance capabilities

of all component materials to various chemical
attacks, especially to water for fibers under stress.

9. stress resistance--the ability of the transmission
line to withstand tensile, torsion and compressive
stresses under dynamic and static conditions and
resistance to impacts and vibrations.

10. radiation resistance--the affecting of fibers by
various sources of nuclear radiation, mainly
causing changes in attenuation and dispersion.

Other types of measurements of optical components are also
important. For example, for graded index fibers changes in the
refractive index profile will reflect in changes in the transmission properties of the fiber. Thus, accurate means of measuring
refractive index profiles is important to maintain high quality
fibers. Some methods include the measurements of diffraction,
interference and near field patterns using light sources and
pattern change measurements to calculate the index profile. One
more popular method is the reflection method where the fiber
cross-section is scanned by a focused laser beam and reflected
laser light intensity is plotted against radial position to calculate the relative refractive index profiles.

The measurement of optical losses and scattering can be
accomplished by first focusing monochromatic light onto a polished
fiber end. The light transmitted into the cladding is removed by
submersing the cladding into an index matched fluid bath. The
scattering of light along the length of the fiber can then be
measured as a radial position function using a photodetector
array surrounding the fiber. To achieve long fiber lengths, the
waveguide can be coiled around a cylinder whose radius exceeds
20cm to minimize radiation effects. Transmitted light received
at the end of the fiber is measured with a photodetector submerged
in core-index matched fluid to minimize reflection losses between
the detector surface and the polished fiber end. Accuracies using
this sort of measurement method can exceed ±0.1dB/Km for fiber

lengths exceeding 300 meters. Overall accuracy increases as longer
fiber lengths are used and when laser light (narrow spectral width
light) is used instead of light generated from LED sources.

For signal distortion measurements, two methods are commonly
used for optical waveguides. One measures the change in the width
of a pulse along a length of fiber to establish a maximum signal
transmission bit rate. The other method measures the baseband frequency response of the fiber to establish a maximum signal bandwidth. The equipment used to measure pulse width widening can
achieve 20ps resolution of the pulse width.

The measurement of the frequency response of the fiber is
accomplished by modulating the optical light source. The transmitted signal is analyzed with a spectrum analyzer for various
fiber lengths, thus establishing the attenuation versus frequency
function for the given waveguide. This type of measurement includes
both material dispersion and group delay distortion thus directly
giving the information bandwidth capability of the transmission
fiber or cable.

In all of these experimental devices, care must be taken
not to induce bend radiation losses by keeping all bending or
coiling of the fibers or cables limited to curvature, radii exceeding 20cm and minimizing the use of any clamping devices requiring large applied stresses to be used on the transmission waveguide.

The mechanical strength testing devices invented for optic
waveguides measure the ability of the fiber or cable to withstand
various sorts of strain by different types of applied stresses.
For cables, the ratings of applied mechanical stress are given by
that required to cause failure in any internal fiber.

The measurement for bending radius ratings is rather simple.
A fiber is wound around a mandrel containing a wide spread of diameters from a few times to several hundred times the fiber diameter.
By testing numerous fibers, an average breaking diameter can be
established at various operational temperatures. For certain

communication systems where the operational temperatures are well
defined, the fiber and cable testing should include tests which
examine the transmission and physical characteristics as a function
of temperature over the known ranges. For general applicability,
the fibers and cables should be tested over a temperature range
with extremes at liquid nitrogen temperatures to the melting point
of the materials used within the cable, usually the plastic
coatings on the fiber or the outer protective jacket. Figure 12
depicts the various mechanical properties testing methods commonly
used for optical fibers.

Tensile and torsional strength are determined by clamping
the ends of the fiber or cable and applying various axial and torsional loads. Increasing loads are used until fracture occurs.
Note that it is possible, especially for torsional loading, for the
cable jacket or internal elements to fail or split before the internal fibers fail. One form of the tensile strength measurement
uses a 90° bend in the fiber or cable where the bend radius is
5-20 times the cable diameter. This test, called the mandrel
strength test, is one simple measure of the cable's ability to
withstand certain field stresses such as those experienced in
stringing telecommunications cables on poles or in the underground
conduit.

The most detailed source of information on fiber measurements is the Military Standards currently used for fiber optic
materials: Fiber Optics Test Methods and Instrumentation MIL-STD-
1678. This document contains a complete listing of definitions,
test requirements, and details of measurement methods and equipment. Measurements and tests included in the military standards
are as follows:

1. fiber size and bundle diameter measurement
2. tensile loading vs. humidity
3. cyclic flexing testing
4. low temperature flexibility--cold bend method
5. impact and twist testing

FIGURE 12

MECHANICAL PROPERTIES TESTING

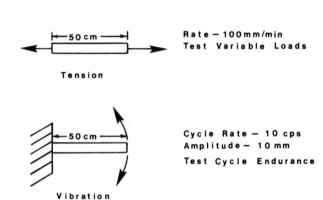

|←— 50 cm —→|
Rate — 100mm/min
Test Variable Loads

Tension

|←— 50 cm —→|
Cycle Rate — 10 cps
Amplitude — 10 mm
Test Cycle Endurance

Vibration

Specify Roller Diameter
(40mm) And Θ (135°)
Go And Return 30 Times
Test Variable Loads (kg)

Jerk-Bend

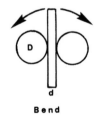

Specify Roller Diameter D
And Cable Diameter d
Go And Return 10 Times
Test Variable D Sizes At
Temperature Extremes

Bend

6. compressive strength testing
7. freezing water immersion--ice crush testing
8. cable tensile loading
9. power transmission vs. temperature and temperature cycling
10. dimensional stability and flammability testing
11. radiant power and radiation pattern measurements
12. attenuation measurements
13. acceptance pattern measurements
14. pulse spreading and crosstalk measurements
15. fiber bundle transfer function measurements
16. refractive index profile (interferometer and near field)
17. insulating blocking, wicking and fluid immersion
18. jacket flow detection and leak testing
19. fiber and bundle and preparation methods.

The details of experimental apparatus and testing methods used in the military specifications should be followed by manufacturers in determining product specifications for publication. This should eliminate some of the confusion originally created in the field when individual companies and universities had no guidelines for reporting test methods and experimental results. Since the field is still in a state of rapid growth and evolution, it will naturally be expected that new and more sophisticated testing methods will be created in the future. It, thus, may be quite some time from now before a complete and final version of testing methods, procedures, and equipment will be available.

IV. Fiber Optic Standards

The International Electotechnical Commission has worked out some international standards to be used for fiber optic communications equipment particularly for international telecommunications. The standards cover the following general areas of both equipment and testing procedures:

1. Mechanical, electrical, and environmental test
   methods, equipment and procedures.
2. Physical and electrical characteristics of fibers
   and cables including light and signal transmission
   characteristics, impurity analytical methods and
   splicing methods.
3. Connectors and couplers for fibers and cables
   including physical size, stress and temperature
   limits, and adhesive quality and ratings.
4. Light sources including LED's and available lasers.
5. Photodetector electrical and mechanical characteristics
   including physical size and environmental limitations.

Manufacturing safety procedures will not be covered
in the standards, but left up to the individual companies and
federal agencies and legislation. Standards for military applications, particularly for aircraft, will be basically the responsi-
bility of the individual armed forces divisions. Some standards,
particularly for electrical components in the presence of com-
bustible fuels, are already covered by existing safety codes.

V. Field Testing

Most equipment used to characterize transmission system
capabilities are designed for laboratory type measurements to
obtain high degrees of accuracy and to cover a variety of anticipated field conditions. However, compact measurement devices
must be made available for field installations, testing, and
repairs. The largest anticipated usage of such instrumentation
would be in telephone transmission equipment and to a lesser extent cable television systems. In both of these cases, a large
number of individual connections and cables must be made over wide
regional areas. Such field-testing equipment could include total
transmission loss measurements and BER's at various locations and
must be able to identify individual fibers in large bundles. The
highest degree of sophistication of field-testing equipment will

probably be required for interactive cable television systems which will be available when economical laser courses are marketed and employed with large bandwidth modulators. In these systems in the eventual evolutionary status, large numbers of commonly used audio and visual channels will be delivered to individual users as well as connecting channels for police alarm systems, station-to-home voting and polling systems, computerized individualized educational communications systems and fire alarm devices.

All field-testing devices will certainly see great evo-lutionary changes in the coming years. At present, the importance of this market in the fiber optics communication area has been somewhat overlooked since the greatest impetus has necessarily concentrated on delivering cables and various optical receivers, transmitters, and modulators.

VI. Power Feed in Cables

Optical fiber cables used for most purposes, such as tele-communications cables, include power transmission wires when re-peaters are used or to drive local closed-loop circuits containing light sources and detector-receivers. Copper wires are usually in-corporated within the outer protective sheath of the cable covered by plastic or paper wrapping. For some cables where metal strength members are used, which includes all outdoor and underwater trans-mission lines, repeaters and components can be powered using the support strands without the addition of internal conducting wire pairs. Using a parallel feed network with uniform metal conduct-ing wires within the optical cable, the basic cost (exclusive of the installation costs) of the repeater or component unit power de-livery is given by: Basic Cost = $24KPL^2NC/V^2$, where K is the weight per unit length - resistance per unit length constant for the metal, P is the power consumed by the individual repeater or component unit, L is the mean distance of the repeater from the feed station or central power unit, N is the total number of re-peaters or component units powered by the transmission line, C is

the cost per unit weight of the plastic coated conducting wires and
V is the feed station delivery voltage. A typical value for K is
154 kg-ohm/km^2 for copper wire. The total required repeater or
component unit power consumption varies widely depending upon such
factors as the transmission bit rate, the number of individual
channels, the use of APD's and lasers versus PIN's and LED's, the
optical repeater spacings and the type of application. Carefully
designed repeaters can have power ratings from 400 milliwatts to
several watts depending upon the application. The value of L depends mostly on N, the desired repeater spacings and the distance
to be transversed between stations or end-terminal devices. Short
hauls will have small values of L while long hauls such as cross-
country telephone lines or transoceanic cables can raise the value
of L well beyond 100km. The cost of coated metal conducting wire
by unit weight typically ranges from $1.40 to $2.00/kg. Copper
wire pairs commonly used in optical cables valued on the 1978 U.S.
market are about $1.60/kg. The voltage range for V can be from
200 to 1000 volts depending upon the repeater power-voltage requirements and the safety break-down limit voltage of the cable-
repeater system under all of the anticipated environmental conditions.

      Most applications will pre-determine the values for P, L
and N by the system requirements of bit rate, terminal spacing and
other factors. Using available commercial wire conductors leaves
V as the only major economic variable usually operating below
700 volts. Clearly, by increasing V to the highest safe level will
decrease or relatively minimize the power delivery (cabling) costs.

      In many fiber optic communications systems repeaters often
incorporate battery power supplies which automatically activate
when the main power delivery wires fail or when the system is re-
wire or undergoing internal repairs. The battery sources can be
designed to last several hours or up to several days depending
upon the weight and space limitations of the repeater unit and the
relative importance of keeping the transmission line in continuous

operation. Once normal power feed stations are resumed the batteries tap sufficient current from the main supply wires to completely recharge. Existing battery packs and trickle-chargers can successfully operate with several hundred breakdowns (activations) and recharging cycles before the battery packs must be replaced. Fiber optic telephone repeaters in Japan have successfully used this type of emergency battery power system to continue transmissions during power failures and expansions of the system services. Automatic cut-in power supplies internal to the transmission link are especially required for military applications and critical process control instrumentation in nuclear and chemical plants.

Other sources of power for optical communications cables are RTG's (radioactive thermoelectric generators) which are self-contained long-life radioactive energy units. These RTG units have a wide variety of designs and internal operating temperatures (400° to 900°C) and have a large range of power delivery capability per individual unit. Those compatible with long distance fiber optic cables require about 1 watt of power for long periods of service. The units must be designed to deliver excess power at the initial stages of installation since the radioactive material heat source will diminish in operating temperature as time passes. Controlled current or voltage devices limit power delivered to the system to maintain a steady drive-current level. As the units age the radioactive sources continually decay causing a reduction in heat generation and thus electrical current. By designing the initial capsule and power limiting devices a wide range of available service times and power capabilities can be tailor-made to suit most applications. The greatest usage of these power supplies will mostly be for long term applications where it is impractical or undesirable to run or to rely on power feed stations and impractical to replace battery operated sources. These applications include certain military installations and

underwater cables at depths below the level at which divers can
safely decend to make repairs to submarine units without the use of
specialized equipment. With the upcoming advances in integrated
optical circuits it is unlikely that underwater cables installed in
the next few years would be repaired to any significant degree.
Rather they will most likely be replaced in ten to twenty years by
high capacity integrated cables which will have reduced power
consumption per channel.

References for Chapter 1

Cables

1. Ando, K. et al. "Performance of Optical Fiber Cables
   Using Plastic Spacers," Second European Conference on
   Optical Fiber Communication, September, 1976, Paris,
   France.

2. Bachel, E. et al. "Jointing Techniques for Optical Cables,"
   ibid.

3. Cook, A. et al. "The Manufacture of Hybrid Multiple Strand
   Optical Fiber Cables," International Conference on Integrated
   Optics and Optical Fiber Communication, July, 1977, Tokyo,
   Japan

4. Ford, S. G. et al. "Principles of Fiber Optical Cable
   Design," Proceedings of IEE, Vol. 123, 1976.

5. Gallawa, R. L., "Considerations in the Use of Optical
   Waveguides in Submarine Cable Systems," OT Report 76-103,
   September, 1976.

6. Gibson, P. T. et al., "Production and Performance of a
   Kevlar-Armored Deep Sea Cable," Proceedings of MTS-IEEE,
   Ocean 1976 Symposium, September, 1976.

7. Gloge, D., "Optical Fiber Packaging and its Influence on
   Fiber Straightness and Loss," Bell System Technical Journal,
   Vol. 54, No. 7, 1975.

8. Hightower, J. D. et al., "Lightweight Cables for Deep
    Tethered Vehicles," Proceedings of MTS-IEEE Ocean 1975
    Symposium, September, 1975.

9. Justice, R. et al. "Designs for Neutrally Buoyant Multi-
    conductor Cables," ibid.

10. Kojima, M. et al., "Flat Type Cable with Silicone Clad Op-
    tical Fibers," International Conference on Integrated Optics
    and Optical Fiber Communication, July, 1977, Tokyo, Japan.

11. Miller, C. M., "Laminated Fiber Ribbon for Optical Communica-
    tions Cables," Bell System Technical Journal, Vol. 55, No. 7,
    1976.

12. Noane, G. L., "Optical Fiber Cable and Splicing Techniques,'
    Second European Conference on Optical Fiber Communication,
    September, 1976, Paris, France.

13. Schwartz, M. I., "Optical Cable Design Associated with
    Splicing Requirements," ibid.

14. Wilkins, G. A., "Fiber Optic Cables for Undersea Communica-
    tions," Fiber and Integrated Optics, Vol. 1, 1977.

15. Wilkins, G. A., "Performance Characteristics of Kevlar-49
    Tension Members," Proceedings of International Conference
    on Composite Materials, Vol. II, April, 1975.

Fibers and Fabrication

16. Abe, K., "Fluorine Doped Silicone for Optical Waveguides,"
    Second European Conference on Optical Fiber Communication,
    September, 1976, Paris, France.

17. Akamatsu, T. et al., "High Deposition Rate CVD Method with
    Helium Gas," ibid.

18. Akamatsu, T. et al., "Fabrication of Long Length Fibers
    by Improved CVD Method," Topical Meeting on Optic Fiber
    Transmission, February, 1977, Williamsburg, Virginia.

19. Albarino, R. V. et al., "An Improved Fabrication Technique
    for Applying Coatings to Optical Fiber Waveguides," ibid.

20. Arima, T. et al., "P_2O_5- GeO_5 - Ga_2O_3 Glass Fibers for
     Optical Communication," International Conference on
     Integrated Optics and Optical Fiber Communication, July,
     1977, Tokyo, Japan.

21. Black, P. W. et al., "Control Techniques for Producing
     Silica Fibers to High Specifications," ibid.

22. Blyler, L. L. et al., "Low-Loss FEP Clad Silica Fibers,"
     Applied Optics, Vol. 14, No. 1, 1975.

23. Bown, T., "Fiber Optics as an Interconnecting Medium,"
     Electronic Packaging and Production, April, 1976.

24. Brehm, G. et al., "Material Optimization and Character-
     ization for Sodalime Silicate Glass Fibers," Second European
     Conference on Optical Fiber Communication, September, 1976,
     Paris, France.

25. Brena, M. et al., "R. F. Induction Furnace for Silica-Fiber
     Drawing," Electronics Letters, Vol. 12, No. 11, 1976.

26. Cohen M. I., "Drawing of Smooth Optical Fibers," Topical
     Meeting on Optic Fiber Transmission, February, 1977,
     Williamsburg, Virginia.

27. DiMorcello, F. B. et al., "Preparation of Low-Loss Optical
     Fibers Using Simultaneous Vapor Phase Deposition and Fusion,"
     Proceedings of the Tenth International Congress on Glass,
     1974.

28. French, W. G. et al., "Single-Mode Fibers with Different
     B_2O_3 - SiO_2 Compositions," Topical Meeting on Optic Fiber
     Transmission, February, 1977, Williamsburg, Virginia.

29. French, W. G. et al., "Low-Loss Optical Waveguides with
     Pure Fused SiO_2 Cores," Proceedings of the IEEE, Vol. 62,
     1974

30. Friebele, E. J. et al., "Drawing Induced Defect Centers in
     Silica Core Fiber Optics," Second European Conference on
     Optical Fiber Communication, September, 1976, Paris, France.

31. Furakawa, M. et al., "New Light Focusing Fibers Made by
     a Continuous Process," Applied Optics, Vol. 13, 1974.

32. Gambling, W. A. et al., "New Silica Based Low-Loss Optical Fiber," Electronics Letters, Vol. 10, 1976.

33. Geittner, P. et al., "Low-Loss Optical Fibers Prepared by Plasma Activated Chemical Vapor Deposition," Applied Physics Letters, Vol. 28, No. 11, 1976.

34. Hammond, C. R. et al., "Optical Fibers Based on Phospho-silicate Glass," Proceedings of IEEE, Vol. 123, 1976.

35. Hosaka, T. et al., "Fabrication of GeO_2-Doped Silica Single Mode Fibers," International Conference on Integrated Optics and Optical Fiber Communication, July, 1977, Tokyo, Japan.

36. Hoshikawa, M. et al., "Step Index Type Optical Fiber Cable," IEE Conference Publication, Vol. 132, 1975.

37. Ikeda, Y. et al., "Development of Low-Loss Glasses for Selfoc Fibers," Second European Conference on Optical Fiber Communication, September, 1976, Paris, France.

38. Inada, K. et al., "Silfa Silicone Clad Optical Fiber," Fujikura Technical Review, No. 2, 1977.

39. Izawa, T. et al., "Continuous Fabrication of High Silica Fiber Preform," International Conference on Integrated Optics and Optical Fiber Communication, July, 1977, Tokyo, Japan.

40. Izawa, T. et al., "Low-Loss Optical Glass Fiber with Al_2O_3 - SiO_2 Core," Electronics Letters, Vol. 10, 1974.

41. Kaiser, P., "Contamination of Furnace-Drawn Silica Fibers," Applied Optics, Vol. 16, No. 3, 1977.

42. Kaminow, I. P. et al., "Ternary Fiber Glass Composition for Minimum Modal Dispersion Over a Range of Wavelengths," Topical Meeting on Optic Fiber Transmission, February, 1977, Williamsburg, Virginia.

43. Kawana, A. et al., "Fabrication of Low-Loss Single-Mode Fibers," Electronics Letters, Vol. 13, No. 7, 1977.

44. Kobayashi, S. et al., "Low-Loss Optical Glass Fiber with Al_2O_3 - SiO_2 Core," Electronics Letters, Vol. 10, 1974.

45. Kuroha, T. et al., "Optical Fiber Drawing and its Influence on Fiber Loss," International Conference on Integrated Optics and Optical Fiber Communication, July, 1977, Tokyo, Japan.

46. Lussier, F. M., "Widening Choices in Fiber Optics, Laser
     Focus, June, 1977.

47. Matsuaka, S., "A Review of Studies on Coating for Optical
     Fibers at Bell Laboratories," Topical Meeting on Optic Fiber
     Transmission, February, 1977, Williamsburg, Virginia.

48. Meyer, F., "The Fabrication of High-Silica Fibers and Multi-
     Component Glass Fibers--A Comparison," Second European Con-
     ference on Optical Fiber Communication, September, 1976,
     Paris, France.

49. Nayyer, J. et al., "Band-Widening of Multimode Optical Fibers
     by Means of Mode Filters with External Higher-Index Sur-
     rounding," Electronics and Communications in Japan, Vol. 58,
     No. 11, 1975.

50. Newns, G. R., "Compound Glasses for Optical Fibers," Second
     European Conference on Optical Fiber Communication,"
     September, 1976, Paris, France.

51. Onada, S. et al., "Frequency Response of Multimode W-Type
     Optical Fibers," Electronics and Communications in Japan,
     Vol. 59, No. 2, 1976.

52. Onada, S. et al., "W-Fiber Design Considerations," Applied
     Optics, Vol. 15, 1976.

53. Rau, K. et al., "Progress in Silica Fibers with Fluorine
     Dopant," Topical Meeting on Optic Fiber Transmission,
     February, 1977, Williamsburg, Virginia.

54. Titchmarsh, J. G., "Fiber Geometry Control with the Double
     Crucible Technique," Second European Conference on Optical
     Fiber Communication, September, 1976, Paris, France.

55. Torza, S., "The Continuous Coating of Glass Fibers," Journal
     of Applied Physics, Vol. 47, No. 9, 1976.

56. Yamazaki, T. et al., "Fabrication of Low-Loss Multicomponent
     Glass Fibers with Graded Index and Pseudo Step-Index Dis-
     tribution," International Conference on Integrated Optics
     and Optional Fiber Communication, July, 1977, Tokyo, Japan.

Fiber Strength

57. Albanno, R. V. et al., "Tensile Strengths of Long Lengths of
    Coated Silica Fibers," Topical Meeting on Optical Fiber Trans-
    mission, February, 1977, Williamsburg, Virginia.

58. Albanno, R. V. et al., "Epoxy-Acrylate-Coated Fused Silica
    Fibers with Tensile Strengths >500ksi in 1 Km Group Lengths,"
    Applied Physics Letters, Vol. 29, 1976.

59. Cress, H. A. et al., "Evaluation of Kevlar Strengthened
    Electromechanical Cable," Proceedings of Materials Technical
    Society Tenth Annual Conference, September, 1974.

60. Evans, A. G. et al., "Proof Testing of Ceramic Materials--
    An Analytical Basis for Failure Prediction," International
    Journal of Fracture, Vol. 10, 1974.

61. Fuller, E. R. et al., "An Error Analysis of Failure Predic-
    tion Techniques Derived from Fracture Mechanics," Journal of
    the American Ceramic Society, Vol. 59, No. 9, 1976.

62. Justice, B., "Strength Considerations of Optical Wavelength
    Fibers," Fiber and Integrated Optics, Vol. 1, 1977.

63. Kalish D. et al., "Probability of Static Fatigue Failure in
    Optical Fibers," Applied Physics Letters, Vol. 28, No. 12,
    1976.

64. Kao, C. K., "Determination of Fatigue Strength and Flow Size
    Distribution for Strong Fibers," Second European Conference
    on Optical Fiber Communication, September, 1976, Paris,
    France.

65. Kao, C. K., "Fatigue Strength of Strong Fibers at High
    Temperatures," Topical Meeting on Optical Fiber Transmission,
    February, 1977, Williamsburg, Virginia.

66. Knapp, R. H., "Nonlinear Analysis of a Helically Armored
    Cable with Non-Uniform Mechanical Properties in Tension and
    Torsion," Proceedings of MTS-IEEE Ocean 1975 Symposium,
    September, 1975.

67. Kobayashi, T. et al., "Tensile Strength of Optical Fiber
     by Furance Drawing Method," International Conference on
     Integrated Optics and Optical Fiber Communication, July,
     1977, Tokyo, Japan.

68. Krause, J. T. et al., "Dynamic and Static Fatigue of High
     Strength Epoxy-Acrylate Coated Fused Silica Fibers," Topical
     Meeting on Optical Fiber Transmission, February, 1977,
     Williamsburg, Virginia.

69. Love, R. E., "Strength of Optical Fibers," SPIE/SPSE Tech-
     nical Symposium, 1976, East Reston, Virginia.

70. Maurer, R. C., "Strength of Fiber Optical Waveguides,"
     Applied Physics Letters, Vol. 27, 1975.

71. Maurer, R. C., "Tensile Strength and Fatigue of Optical
     Fibers," Journal of Applied Physics, Vol. 47, No. 10, 1976.

Measurements

72. Adams, M. J. et al., "Determination of Optical Fiber Refrac-
     tive Index Profiles by a Near-Field Scanning Technique,"
     Applied Physics Letters, Vol. 28, 1975.

73. Adams, M. J. et al., "Measurement of Profile Dispersion
     in Optical Fibers: A Direct Technique," Electronics Letters,
     Vol. 13, No. 7, 1977.

74. Akiba, S. et al., "Carrier Lifetime Measurement of InGaAsP/
     InP DH Lasers," International Conference on Integrated Optics
     and Optical Fiber Communications, July, 1977, Tokyo, Japan.

75. Arnaud, J. A. et al., "Novel Technique for Measuring the
     Index Profile of Optical Fibers," Bell System Technical
     Journal, Vol. 55, No. 10, 1976.

76. Ashe, H. W. et al., "Frequency Domain Measurements of Dis-
     persion as a Function of Wavelength in Multimode Optical
     Fibers," Topical Meeting on Optical Fiber Transmission,
     February, 1977, Williamsburg, Virginia.

77. Bouillie, R. et al., "Measurement Methods of Optical Conductor Transmission Characteristics," Second European Conference on Optical Fiber Communication, September, 1976, Paris, France.

78. Brandt, G. B., "Two Wavelength Measurements of Optical Waveguide Parameters," Applied Optics, Vol. 14, No. 4, 1975.

79. Cannell, G. J., "Continuous Measurement of Optical Fiber Attenuation During Manufacture," Electronics Letters, Vol. 13, No. 5, 1977.

80. Chu, P. L., "Determination of the Diameter of Unclad Optical Fiber," Electronics Letters, Vol. 12, No. 1, 1976.

81. Cohen, L. G., "Pulse Transmission Measurements for Determining Near Optical Profile Gradings in Multigrade Borosilicate Optical Fibers," Applied Optics, Vol. 15, No. 7, 1976.

82. Cohen, L. G. et al., "Wavelength Dependence of Frequency Response Measurements in Multimode Optical Fibers," Bell System Technical Journal, Vol. 55, No. 10, 1976.

83. Costa, B. et al., "Measurements of the Refractive Index Profile in Optical Fibers: Comparison Between Different Techniques," Second European Conference on Optical Fiber Communication, September, 1976, Paris, France.

84. Eickhoff, W. et al., "Measuring Method for the Refractive Index Profile of Optic Glass Fibers," Optical and Quantum Electronics, Vol 7, No. 2, 1975.

85. Euler, F. K. et al., "Measurements of Layer Thickness and Refractive Indices in High Energy Ion-Implanted GaAs and GaP," Journal of Applied Physics, Vol. 47, No. 12, 1976.

86. Friebele, E. J. et al., "In Situ Measurement of Growth and Decay of Radiation Damage in Fiber Optic Waveguides," Topical Meeting on Optical Fiber Transmission, February, 1977, Williamsburg, Virginia.

87. Gloge, D. et al., "Profile Dispersion in Multimode Fibers: Measurement and Analysis," Electronics Letters, Vol. 11, 1975.

88. Hara, E. H. et al., "Measurement of Non-Linear Distortion
     in Light-Emitting Diodes," Electronics Letters, Vol. 12, 1976.

89. Hendirich, P. F. et al., "Optical Waveguide Refractive Index
     Profiles Determined from Measurement of Mode Indices: A
     Simple Analysis," Applied Optics, Vol. 15, No. 1, 1976.

90. Heitman, W., "Attenuation Measurement in Glasses for Optical
     Communications: An Immersion Method," Applied Optics, Vol. 15,
     No. 1, 1976.

91. Hotate, K. et al., "Refractive Index Profile of an Optical
     Fiber: Its Measurement by the Scattering-Patten ·Method,"
     Applied Optics, Vol. 15, No. 11, 1976.

92. Iga, K. et al., "Precise Measurement of the Refractive Index
     Profile of Optical Fibers by Non-Destructive Interference
     Method," International Conference on Integrated Optics and
     Optical Fiber Communications, July, 1977, Tokyo, Japan.

93. Kaiser, P., "Numerical Aperture Dependent Spectral-Loss
     Measurements of Optical Fibers, Ibid.

94. Kaminow, I. P. et al., "Binary Silica Optical Fibers: Refrac-
     tive Index and Profile Dispersion Measurements," Applied
     Optics, Vol. 15, No. 12, 1976.

95. Keck, D. B., "Spatial and Temporal Power Transfer Measure-
     ments on Low-Loss Optical Waveguide," Applied Optics,
     Vol. 13, 1974.

96. Keck, D. B. et al., "Measurement of Differential Mode
     Attenuation in Graded-Index Fiber Optical Waveguides,"
     Topical Meeting on Optical Fiber Transmission, February,
     1977, Williamsburg, Virginia.

97. Marcatili, E. A. J., "Factors Affecting Practical Attenua-
     tion and Dispersion Measurements," ibid.

98. Marcuse, D. et al., "Measurement of Fiber Coating Concen-
     tricity," ibid.

99. Mims, F. M., "Measuring LED Power Distribution," Electro-
     Optical Systems Design, June, 1976.

100. Presby, H. M., "Ellipticity Measurement of Optical Fibers,"
     Applied Optics, Vol. 15, No. 2, 1976.

101. Ramaswamy, V. et al., "A New Method for Measuring Parallel Waveguide Directional Coupler Parameters," Topical Meeting on Integrated Optics, January, 1976, Salt Lake City, Utah.

102. SeiRai, S. et al., "Measurement of Baseband Frequency Response of Multimode Fiber Using New Type of Mode Scrambler," Electronics Letters, Vol. 13, No. 5, 1977.

Transmission Properties and Losses

103. Aritone, H. et al., "Propagation Characteristics of Graded-Index Optical Waveguides Fabricated by Ion Implantation," Topical Meeting on Integrated Optics, January, 1976, Salt Lake City, Utah.

104. Arnaud, J. A., "Pulse Broadening in Multimode Optical Fibers," Bell System Technical Journal, Vol. 54, No. 7, 1975.

105. Arnaud, J. A., "Effect of Polarization on Pulse Broadening in Multimode Graded Index Optical Fibers," Electronics Letters, Vol. 11, 1975.

106. Bodem, F. et al., "Investigation of Various Transmission Properties and Launching Techniques of Plastic Optical Fibers Suitable for Transmission of High Optical Power," Optical and Quantum Electronics, Vol. 7, No. 5, 1975.

107. Buckler, M. J. et al., "Optical Fiber Transmission Properties Before and After Cable Manufacture," Topical Meeting on Optical Fiber Transmission, February, 1977, Williamsburg, Virginia.

108. Byron, K. C. et al., "The Impulse Response of Optical Fibers," Second European Conference on Optical Fiber Communication, September, 1976, Paris, France.

109. Carnstam, B. et al., "Attenuation in Bends of Diffused Waveguides," Topical Meeting on Integrated Optics, January, 1976, Slat Lake City, Utah.

110. Cherm, A. H. et al., "Delay Distortion Characteristics of Optical Fiber Splices," Applied Optics, Vol. 16, No. 2, 1977.

111. Cohen, L. G. et al., "Length Dependence of Pulse Dispersion
      in a Long Multimode Optical Fiber," Applied Optics, Vol. 14,
      No. 6, 1975.

112. Cohen, L. G. et al., "Pulse Dispersion in Multimode Fibers
      with Graded B_2O_3 - SiO_2 Cores and Uniform B_2O_3 - SiO_2
      Cladding," Applied Physics Letters, Vol. 28, 1976.

113. French, W. G. et al., "Polarization Effects in Short
      Length Single-Mode Fibers," Topical Meeting on Optical
      Fiber Transmission, February, 1977, Williamsburg, Virginia.

114. Gardner, W. B., "Microbending Loss in Optical Fibers,"
      Bell System Technical Journal, Vol. 54, No. 1, 1975.

115. Hoshikawa, M. et al., "Transmission Properties of Ring-Type
      Optical Fibers," Second European Conference on Optical
      Fiber Communication, September, 1976, Paris, France.

116. Howard, A. Q., "Bend Radiation in Optical Fibers," Fiber
      and Integrated Optics, Vol. 1, 1977.

117. Ikeda, M. et al., "Pulse Separating in Transmission Charac-
      teristics of Multimode Graded Index Optical Fibers," Applied
      Optics, Vol. 15, No. 5, 1976.

118. Inada, K. et al., "Experimental Consideration of the Pulse
      Spreading of Multi-mode Optical Fibers", Electronics Commun-
      ication Engineers of Japan, Vol. 75, 1975.

119. Inada, K., et al., "Losses Due to Core-Cladding Interface
      Imperfection and Their Reduction in Optical Fibers Made
      by CVD Technique," Topical Meeting on Fiber Optic Trans-
      mission, February, 1977, Williamsburg, Virginia.

120. Ishikawa, R. et al., "Transmission Characteristics of
      Graded Index and Pseudo-Step Index Borosilicate Compound
      Glass Fibers," International Conference on Integrated Optics
      and Optical Fiber Communication, July, 1977, Tokyo, Japan.

121. Jeuhomme, L. et al., "Mode Coupling in a Multimode Optical
      Fiber with Microbends," Applied Optics, Vol. 14, No. 10,
      1975.

122. Jurgensen, K., "Dispersion-Optimized Optical Single Mode
     Glass Fiber Waveguide," Applied Optics, Vol. 14, No. 1,
     1975.

123. Kajoka, H. et al., "The Influence of Mechanical Stress on
     Transmission Characteristics of W-Type Optical Fibers,"
     Second European Conference on Optical Fiber Communication,
     September, 1976, Paris, France.

124. Kapron, F. P., "Maximum Information Capacity of Fiber Optic
     Waveguides," Electronics Letters, Vol. 13, No. 3, 1977.

125. Kawakami, S. et al., "Anomalous Dispersion of New Doubly
     Clad Optical Fiber," Electronics Letters, Vol. 10, 1974.

126. Keck, D. B. et al., "Pulse Broadening in Graded-Index
     Optical Fibers," Applied Optics, Vol. 5, No. 2, 1976.

127. Kobayashi, I. et al., "Transmission Characteristics of a
     Graded-Index Multimode Fiber," International Conference on
     Integrated Optics and Optical Fiber Communication, July,
     1977, Tokyo, Japan.

128. Lewin, L., "Radiation from Curved Dielectric Slabs and
     Fibers," IEEE Transacitons on Microwave Theory and Tech-
     niques, MTT-22, 1974.

129. Marcatili, E. A. J., "Modal Dispersion in Optical Fibers
     with Arbitrary Numerical Aperture and Profile Dispersion,"
     Bell System Technical Journal, Vol. 56, No. 1, 1977

130. Marcuse, D., "Scattering and Absorption Losses of Multi-
     mode Optical Fibers and Fiber Lasers," Bell System Technical
     Journal, Vol. 55, No. 10, 1976.

131. Miyagi, M. et al., "Mode Conversion and Radiation Losses
     in a Step-Index Optical Fiber Due to Bending," Optical
     and Quantum Electronics, Vol. 9, No. 1, 1977.

132. Olshansky, R., "Model of Distortion Losses in Cables
     Optical Fibers," Applied Optics, Vol. 14, 1975.

133. Pask, C. et al., "Multimode Optical Fibers: Interplay of
     Absorption and Radiation Losses," Applied Optics, Vol. 15,
     No. 5, 1976.

134. Petermann, K., "Pulse Distortion in Single Mode Fibers
      with Microbending," Electronics Letters, Vol. 13, No. 2,
      1977.

136. Rokunohe, M. et al., "Stability of Transmission Properties
      of Optical Fiber Cables," Second European Conference on
      Optical Fiber Communication, September, 1976.

136. Suematsu, Y. et al., "Transmission Characteristics of
      Mode-Coupled Multimode Optical Fibers," International
      Conference on Integrated Optics and Optical Fiber Com-
      munication, July, 1977, Tokyo, Japan.

137. Tanaka, T. et al., "Frequency Response of Multimode W-Type
      Optical Fibers," Electronics Communications Engineers of
      Japan, Vol. 75, 1975.

CHAPTER 2

COUPLERS, CONNECTORS, AND SPLICES

I. Introduction

Fiber optic communications systems contain a number of individual components, fibers, light sources, detectors, repeaters, end devices, etc., which must be efficiently linked together to make the optical system competitive with the existing electrical communications systems. Each type of link, fiber-to-fiber, fiber-to-detector, fiber-to-source, fiber bundle-to-fiber bundle, cable-to-repeater, etc., has unique engineering problems to overcome to achieve efficient couplings of high physical integrity. The splices, couplers and connectors necessary for these links are described as follows.

A splice is a form of a coupler that permanently joins two fibers or two fiber bundles. A coupler links two or more fibers together providing two or more paths for the transmission signal. A coupler is active if it provides a switching mechanism for route selection and passive if its routing is fixed by the geometry of the coupler. A connector links one fiber to another or to repeaters or end-devices such that as much as possible of the original signal is transmitted into the second fiber, fiber bundle or end device. Connectors are usually demountable from the rest of the fiber optic transmission system. These demountable connectors have some sort of locking mechanism, and must introduce a minimal amount of insertion loss with reproducibility when demounting and reinserting into the transmission path.

All splices, couplers and connectors need optically flat fiber ends which must remain clean during installation and maintenance work where connections may be opened and closed a number of times. These linking mechanisms must also be available at low cost and be reasonably simple to install for field applications. Some of the problems that must be dealt with to achieve low cost, high strength, low loss and reproducible couplers, splices and connectors are that many fibers are thin and fragile with small coupling areas and that fibers and cables cover a wide variety of sizes, shapes and materials of construction.

The splicing and connecting of optical fibers can be made with either permanent or detachable connections. Underwater cables employ permanent connections. Where upgrading of the communications system is expected, detachable connections should be made on repeater-modulator assemblies to make equipment replacement easier in the future. In some applications the size and weight of the linking devices must be small in size to fit into small or crowded spaces. All types of connectors should be easy to work with in the field both in terms of coupling together factory installed connectors and for maintenance and repair work. All losses incurred in linking fiber optic components should be minimized, especially in large bandwidth systems and where numerous connections are expected between the light source and end-device.

II. Fiber Splicing

Fiber splices for high data rate transmission lines must have low insertion losses, usually expected to be less than 0.6dB, and have sufficient mechanical strength compared to the rest of the fiber or cable such that it is not a weak link in the line. Numerous types of splices are available for fiber splicing. Figure 13 depicts some of the various splicing methods. The procedures used for splicing depend on the type of fibers and cables to be joined. For example the splices used for large underwater cables are different in size and structure than those used for indoor ribbon

FIGURE 13

VARIOUS SPLICING METHODS

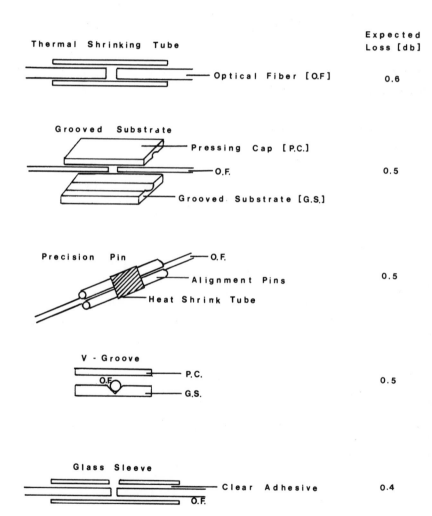

Expected
Loss [db]

Thermal Shrinking Tube

Optical Fiber [O.F]   0.6

Grooved Substrate

Pressing Cap [P.C.]
O.F.   0.5
Grooved Substrate [G.S.]

Precision Pin — O.F.

Alignment Pins   0.5
Heat Shrink Tube

V - Groove

P.C.
O.F.   0.5
G.S.

Glass Sleeve

Clear Adhesive   0.4
O.F.

splices.

The ends of the optical fibers must first be prepared before they are spliced together. The ends must be clean, flat and perpendicular to the axis of the fiber. Fibers may be cleaned with smooth surfaces by scribing with a small diamond edge while the fiber is kept under tension. The surfaces prepared in this manner do not usually have to be processed further by grinding or polishing. Various score-and-flex methods have been devised for single and ribbon fibers.

The procedure for most fibers and cable splices are as follows.

1. Remove the protective coatings from the cable (sheath) and fibers (plastic coatings) to expose the inner optical fibers.

2. If strength members are used in the cable they must be suitably mated with wire or plastic filament couplers, to maintain the strength integrity of the cable.

3. The fiber ends are prepared by scoring and breaking or by some other method such as cutting, grinding and polishing, to give flat clean surfaces.

4. The fibers to be jointed must be first identified (which may be burdensome for multifiber non-color coded cables) and then inserted into the chosen splicing device such as a small glass tube or grooved substrate splice.

5. The fibers are then bonded together with a high strength epoxy resin or other suitable adhesive that refractive index-matches the fibers. (Fibers may also be fused thermally by various techniques.)

6. When the splice has been completed the protective coating to the fiber and protective covering on

the cable must be reapplied. (For some applications tapes and cushionings can be added around
the splice to protect it from external impacts
and stresses.

Two common field splicing techniques are the precision
sleeve and loose tube splice. The precision sleeve splice requires
high degrees of accuracy and reproducibility of the inner sleeve
dimensions and on the outer fiber diameter. For splices using
index-matching fluids and adhesives, precision tube splices have
consistently achieved insertion losses less than 0.3dB. Some problems occasionally arise with this method when particles of dirt or
glass are scraped off the inside wall of the precision glass sleeve
and become trapped between the fiber ends, causing end separation
losses and slight radial losses. The loose tube splice uses a
glass tube with a square cross-section where the inner tube dimensions are purposely made larger than the outer fiber diameter.
When the two fibers are bowed slightly the loose tube rotates and
holds the ends in place until the adhesive has cured. Various modi-
fications of the loose tube splice have been made with losses ranging from 0.3 to 0.5dB. In all of these splicing techniques suitable tools have already been made available to the consumer to
prepare the fiber ends, align the fibers, inject or apply the adhesives and recoat the fibers and cables.

Precision pin splices have also been used with reasonable
success. They are practical in that accurate steel pins with $\pm.4\mu m$
diameters are available commercially. The cleaned pins are clus-
tered around the fibers to be spliced together for accurate alignment and then held in place by thermo-plastic heat-shrink tubes.
Clear epoxy adhesive which matches the index of refraction of the
fibers can then be applied to the inside of the joint using micro-
syringes or wipe applicators. For multi-single fiber cables, precision pin splices may present a space problem in that the total
volume occupied by this splice is larger than the tube splices
which can be a significant factor when many fibers must be joined

together in a small space. Precision pin splices may also be a disadvantage in areas where no metals can be used, since the pins are
usually steel.

Several successful techniques have been developed using fiber welding. Fiber welding is accomplished using precision laser
or electric-arc welding devices and directly heating the fiber ends
or by adding an intermediate low temperature melting glass between
the fiber to minimize induced strain and weak spots. Fiber welding
often creates internal strain within the fiber and although transmission efficiencies can be as high as other methods it is not preferred since it weakens the overall fiber strength and thus reduces
the expected fiber life-time-in-use. Reductions in tension limits
by fusion splicing are about two-thirds of the original fiber stress
rating. Insertion losses ranging from 0.2 to 0.5dB can be achieved
in the laboratory rather consistently. The problem with using
these types of splices is that they are not suited to many types
of field applications both on the basis of portability and potential flammability hazards in certain installation environments.

In addition to simple attenuation, insertion losses in
splices misalignments cause the elimination of certain transmission
modes resulting in an increased dispersion across the splice.
Without any splicing, the dispersion of most fibers ranges from
0.3 to 1 ns/Km. Most splices introduce dispersions ranging from
100 to 400 ps/Km. When large numbers of splices are used these
dispersion losses can rapidly accumulate and in some cases may be
the limiting factor in repeater spacing or critical transmission
length.

Alignment techniques, devices, materials and hardware are in
general advanced enough to achieve consistent splices with losses
less than 0.3dB. The major concern to minimize losses then turns
to the matching of fiber numerical apertures, the fiber core index
profile width and shape, and tolerances on fiber outside diameter
and concentricity. When designing an optimal fiber communications
system the fiber, cable and splicing fabrication methods and designs

must all be considered as inter-relating to keep attenuation to a
minimum.

As with many aspects of fiber communication systems little is
known about the long-term stability of splices and couplings especially in some harsh field environments such as exposed telephone
lines and underwater cables. Splices most likely to obtain extensive field usage will be those using permanent clear adhesives and
those splices which do not use metal alignment components. For applications which can expect to see wide swings in temperature,
splices must also be field designed to have matching or stress
tolerant coefficients of thermal expansion compared to the fibers
which are relatively inelastic. Splices which will undergo extremely high temperature extremes have strict materials limitations
and present coupling problems which have not been carefully studied
for long-term field applications. Telephone line splices for example in northern climates will have to be designed to withstand
temperatures ranging from -50°C to 50°C without stress failure.

III. Alignment Losses

The proper physical alignment of fibers is important to minimize coupling losses. In joining two fibers, a light source to a
fiber or a detector to a fiber, the alignment can be described in
terms of radial (lateral), end (longitudinal) and angular misalignments or displacements. In coupling a light source to a fiber the
longitudinal and angular misalignments have little effect in coupling efficiency as long as reflection losses are minimized. However, due to the small surface area of the fiber core, lateral or
radial misalignments can have significant impact on coupling losses.
An end displacement of 50 μm incurs a loss of about 4dB for a typical fiber-LED coupling where a radial displacement of 50 μm may result in a coupling loss of more than 8dB.

The expected insertion losses in coupling two fibers depends
on whether the refractive index profile is a step or graded index
type. Experimental measurements of the three types of misalignments

are usually reported by using dimensionless lateral and longitudinal distances by dividing the misalignment distance by the fiber diameter. Figure 14 shows typical longitudinal and lateral alignment splicing losses. Figure 15 shows typical angular displacement losses. The losses incurred by end and radial displacements can be minimized by using index-matching fluids or cement between the fibers which should be kept as close together as possible. The losses incurred by such misalignments are also diminished by using large diameter fibers and precision alignment and splicing devices.

IV. Multiple Single Fiber Splicing

The methods commonly used to splice single fiber or several fibers are not really feasible for multiple fiber cables, both in terms of economics and physical restraints. The devices and materials for multiple fiber splicing are not as advanced as single fiber splices but some excellent devices and techniques have been tried for certain fiber systems. The multiple grooved substrate splice works well for ribbon cables, especially when an injector hole is provided for in the cover plate for clear adhesives, and can make splices with losses of 0.2 to 0.4dB.

Special designs have been made specifically for certain multi-fiber cables. In these cases the manufacturer of the cable terminates the cable with a multi-hole or multi-grooved disc or cylinder inserted in the cabling machine before the fibers are brought closely together. Once the fibers are pulled closely together they are glued to the disc or cylinder which is subsequently cut perpendicular to the fiber axis and the ends polished ready for splicing. The discs of cylinders are aligned with interlocking pins and clear adhesive is used in between the splice to reduce reflection losses.

V. Couplers and Connectors

The three general requirements for coupling are as follows. First, the light source must be coupled to the fiber to transmit as

FIGURE 14

LINEAR ALIGNMENT SPLICING LOSSES

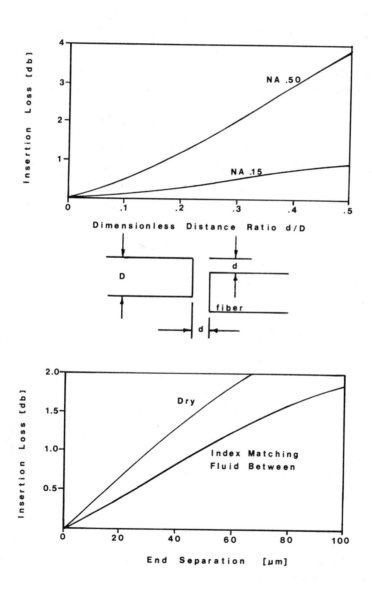

FIGURE 15

ANGULAR ALIGNMENT SPLICING LOSSES

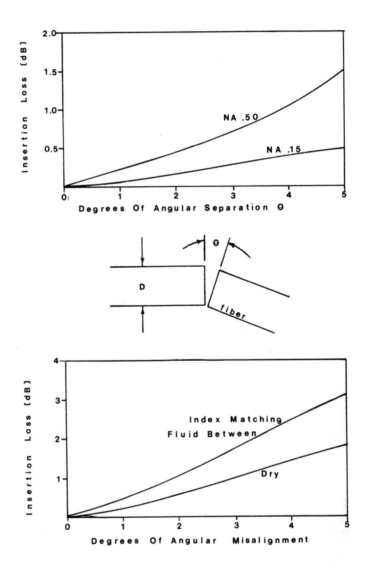

much energy as physically possible into the desired signal format. Second, the transmitted optical energy must be coupled as efficiently as possible from the waveguide to the detector, repeater, end-device or signal-splitting device. Third, the light energy must be coupled from one fiber to another or from one fiber bundle to another as efficiently as possible. All of these couplings should be accomplished with the minimum amount of losses and should not induce reflected energy back into the waveguide.

Connectors are used to join fiber ends in a make-and-break fashion and thus cannot use index matching fluids to minimize reflection losses. It is important to maintain fiber core alignment after each uncoupling and coupling. This is accomplished in most connectors by using one surface (shoulder) of the connector for longitudinal or end spacing and two surfaces for lateral or radial alignments. In dusty environments where particulate matter is a concern connectors should be designed to keep foreign matter from getting to the inside on the fiber ends. Removal of particulates can be accomplished in a manner similar to dusting off delicate coated lenses by using dry compressed gases sprayed over the surfaces before reconnecting the fibers. The milling or molding tolerances in the pieces (sleeve, threads, cylinders, etc.) used to make a connector are very stringent since misalignments of several μm or several degrees can add to insertion losses. Regardless of connector precision, fibers have minor geometric variations which add to potential misalignments even for perfectly mated male-female connector pieces. For step index fibers the longitudinal and lateral alignments are not as critical as those for graded index fibers. Commercially produced connectors have insertion losses ranging from 1.0 to 2.5dB for single and ribbon fibers and for multifiber bundles, with costs ranging from $3 for cheap bundle connectors to $100 for precision single fiber connectors.

Table VI lists some manufacturers and suppliers of fiber optic connectors and splices.

Table VI

Some Manufacturers and Suppliers
of Fiber Optic Connectors or Splices

AMP	Hellerman-Deutsch
Aoi Sansho	ITT
Bell-Northern	Meret
Belling & Lee	NEC
Bunker Ramo	Opto Micron Industries
Deutsch	Sealectro
Electro-Fiberoptics	Spectronics
Fujitsu	Thomas & Betts Corp.
Furukawa Electric	Thomson & CSF

VI. Multi-Fiber Bundle Connectors

The physical tolerance and alignment requirements of multi-fiber bundle connectors are less stringent than those for single fiber connectors, particularly since bundles have large effective NA values. The insertion loss for fiber bundles with relatively large numbers of individual fibers is insensitive to the rotational positions of the transmitting and receiving ends of the coupling. Most fiber bundle connectors mimic typical electronic connectors and are thus simple in design and inexpensive to manufacture. Insertion losses ranging from 2.5 to 4.5dB can be expected depending mostly upon fiber type number and diameter. Larger diameter fibers with thick coatings have higher insertion losses due to an increase in fiber area mismatching between the connector ends. To reduce insertion losses with multi-fiber bundle connectors the fiber coatings should be removed at the last few mm of the fibers and the

fiber diameters should be kept reasonably small to reduce packing
fraction losses.

Laboratory multi-fiber bundle connectors have been made with
insertion losses of less than 1.5dB. Production connectors have not
yet achieved such low values. The average cost in 1977-78 of typical production multi-fiber bundle connectors is from $3 to $10. By
comparison low loss connectors designed for use with index matching
fluids range from $15 to $20, depending upon their insertion loss
rating and the diameter of the bundle to be coupled. For signal
splitting, using multiple-fiber bundle multi-part couplers, several
tee and star couplers are available for limited applications.
Simple tee couplers have insertion losses of about 2-3dB while star
coupler losses range from 5-7dB depending upon the number of interfaces used, the NA values of the components and the losses incurred
in the scramblers. The magnitude of the insertion losses also depends upon the total number of fibers entering and leaving the
couplers and the multi-fiber bundle or multi-single fiber cable diameter.

VII. Light Source Coupling

LED's are available in a large variety of geometric configurations and may be categorized into two basic types: small area-
high brightness and large area-low brightness. The typical low
brightness LED's have emission areas ranging from 0.2 to 5.6mm^2.
The core diameters of typical low loss fibers range from 50 to
100 μm, equivalent to 0.002 to 0.008mm^2. The mismatch in area of
more than two orders of magnitude between fiber and light source
results in very large coupling losses making low brightness-large
area LED's impractical to use for single fibers. In cables containing fiber bundles of 50 to 1000 fibers, these emitters can be used
with reasonable efficiency since the total fiber bundle cross-
sectional area is close to that of the LED.

High brightness-small area LED's and injection lasers with
small emitting surfaces can be coupled rather efficiently to

optical fibers. The Burrus type LED can be directly attached to
the fiber end using a suitable adhesive. Figure 16 shows the
cross-section of a typical Burrus LED coupled to a fiber. The
light emitted from the active surface area of the LED has a characteristic angular distribution (brightness or intensity value)
which can be given by:

$$I_e(\theta) = I_e(\cos \theta)^n \quad ; \quad I_e = \left(\frac{n_m}{n_{LED}}\right)^2 I_d / A_{LED}$$

where I_e is the intensity of the emitted light as a function of
the angular distribution, n is the angular exponential dependence
constant, n_{LED} is the refractive index of the LED interface, n_m is
the refractive index of the surrounding medium, I_d is a constant
depending upon the LED drive current and A_{LED} is the total surface
emitting area of the LED. The manufactures of LED's typically
give the angular radiation pattern as part of the product description. The pattern from edge-emitting DH LED's is more directional
in the sense of having a tighter angular pattern. The most narrow
patterns or most directional are laser diodes. Figure 17 gives
two typical light intensity curves as a function of angular distribution for light sources.

The amount of energy that can be coupled into a fiber is dependent upon its NA as well as the angular directional characteristics of the light source. An optical waveguide can accept only
those light rays which are less than the maximum angle determined
by total internal reflection at the cladding-core interface. Rather
substantial coupling losses can be incurred when the angular distribution of the light source exceeds the angular acceptance confined
by the fiber numerical aperture.

It should be noted that although the typical LED emits a
Lambertian angular distribution, sometimes an off-axis halo surrounding the central beam is created by internal reflections from
the interior of the can used to house the LED chip or external lens.
If improperly designed this off-axis lobe can contain a significant

FIGURE 16

HIGH RADIANCE BURRUS TYPE LED

COUPLED TO OPTICAL FIBER

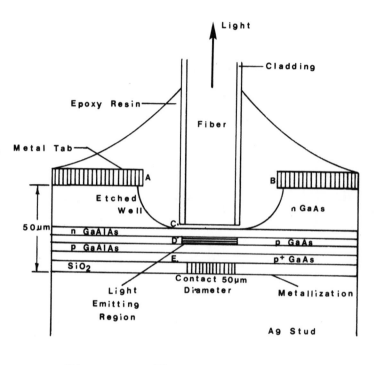

$\overline{AB}$ – 1.5 mm $\overline{CD}$ – 13 μm $\overline{DE}$ – 4 μm

FIGURE 17

LIGHT INTENSITY AS A FUNCTION

OF THE ANGULAR DISTRIBUTION

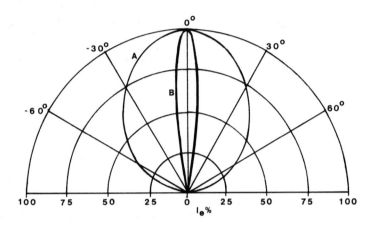

Typical LED Power Curves

A — Emitters With Planar Windows

B — Emitters With Lens Attached

fraction of the total power actually emitted by the LED.

VIII. Coupling Efficiency

The coupling between LED's, lasers and photodetectors to optical fibers can be described in terms of transmitted power by:

$$P_f = F_p \left(\frac{n+1}{2} \right) \left(\frac{A_f}{A_s} \right) (NA)^2 \, P_s,$$

where P_f is the optical power coupled into or out of the fiber, P_s
is the power emitted by the source, F_p is the packing fraction, A_f
is the fiber core area, A_s is the active area of the source or

detector (if $A_f < A_s$, $A_f/A_s = 1$) and n is the angular distribution exponent $(\cos\theta)^n$. NA values typically range from 0.15 to 0.50 with n
values for LED's usually equal 1 and for lasers and LED's with
built in lenses n can equal 2 to 4, since injection lasers in particular have a narrower directional beam. For both step and graded
index fibers, dispersion increases with increasing values of NA
which limits the practical power coupling efficiency by attempting
to use high values of the fiber NA.

The packing fraction for a hexagon close-packed bundle is
given by:

$$F_p = \left(\frac{N}{n_d^2}\right)\left(\frac{d_{core}}{d_{clad}}\right)^2$$

where N is the total number of fibers in the bundle, n_d is the number of fibers along the enclosing diameter, d_{core} is the fiber
core diameter, and d_{clad} is the diameter of the fiber cladding. To
achieve high values of F_p for optimal coupling of light into the
fiber transmission line the value of d_{core} must approach that of
d_{clad}. For $d_{core} \simeq d_{clad}$ the cladding must be stripped from the
core. For typical fiber bundles with hexagonal close-packing the
packing fraction loss ranges from 1.5 to 6.0dB depending upon the
number of fibers in the bundles and whether or not the cladding is
removed from the fiber ends to enhance coupling efficiency.

The packing fraction for a linear close-packed bundle is
given by:

$$F_p = \frac{.25\pi N}{1 + (N-1)\left(\dfrac{d_{clad}}{d_{core}}\right)}$$

For linear .fiber configurations when $d_{clad} \simeq d_{core}$ by stripping
the cladding off the core, the packing fraction becomes simply $\pi/4$,
independent of the total number of fibers. For typical linear ar-
rangements the pack fraction loss ranges from 1 to 3dB depending
upon the core and cladding diameter ratios. Although linear bundles (ribbons) have a lower packing fraction value, in practice it

is difficult to design rectangular couplers and lenses to take full advantage of the minimal coupling losses. In addition, for any type of bundle configuration if too much of the cladding is removed from the fiber, losses will be sustained along the length of the fiber where the cladding has been reduced in diameter or totally removed. Thus, an approximate optimal value of the cladding diameter at the end of the fiber exists for each fiber bundle configuration dependent in part on the physical dimensions of the couplers.

Coupling efficiency between light sources and detectors to fibers can improve dramatically by using some form of transfer optics usually a magnifying lens between the LED, injection laser or photodetector and the fiber end. The purpose of the lens in front of the fiber end is to make the effective NA of the source and fiber approximately equal. Assuming that no losses occur in the lens or in the spaces between source and fiber, lens and fiber or detector and fiber the lens can theoretically project all of the source emitted beam to the fiber end. In practice if air gaps exist between the three basic elements, reflection losses are incurred. In addition, if the lens is not carefully chosen to have high transmission at the wavelength or wavelength region of the emitted light, absorption and scattering losses will occur inside the lens surfaces.

Matching problems also arise in using lenses to collect emitted light. If the light source area is relatively large the fiber core diameter must be proportionally larger in a ratio given by the emission angle of the source divided by the acceptance angle of the fiber. A typical value for this ratio is about 5/1. Thus, for a source area of 20 µm the fiber core would have to be a minimum of 100 µm to achieve optimal lens coupling. Reflection losses between the light source and lens can be minimized by employing index matching epoxies between the emitter surface and the lens and between the lens and fiber. The NA mismatch between light source and fiber is the predominant cause of the insertion loss coupling. Insertion losses from 10 - 30dB are common to many LED to fiber couplings. For laser to fiber couplings the insertion losses are

usually lower since the emission beam is highly directional along
the fiber axis and the emission areas are very small. Using a suitable lens system with index-matching fluid or thin-film coatings on
the laser and fiber to reduce reflection losses insertion losses of
only 2dB and lower can be achieved.

The coupling of photodetectors to fiber ends is more easily
accomplished than with light sources, since they have relatively
large active surface areas and large acceptance angles. Some unique
problems exist for photodetectors, however, such as total elimination
of undesired radiation and point-to-point response variations over
the active area which can induce extraneous signal modulation or
noise by small mechanical movements or vibrations of the detector
with respect to the fiber axis. By using index matching fluids or
thin films, insertion losses with detectors of less than 1dB can be
obtained. Multimode fibers are superior to single mode fibers in
terms of efficiency of coupling light energy from the source to the
fiber.

IX. Integrated Optical Circuits

The applications of integrated optical circuits have been
mostly limited to laboratory experiments for several reasons.
First, the coupling of optical power into and out of an IOC is rather inefficient particularly if one mode structure must be maintained during transmission. Second, the designs and mass fabrication technology of long life-time room temperature solid-state
lasers suitable for the integrated optical circuit format are not
yet available. Third, the applications of integrated optical circuit systems are for wide bandwidth (>1 GHz) communications links
using single-mode fibers, which has limited commercial application
and thus limited production and economic incentive.

A total integrated optical circuit communications system when
available would have the advantages of the cost reduction of opto-
electronic interfaces and the achievement of miniature, power-saving,
rugged component assemblies. The sophistication in technology to go

from an opto-electronic system to an IOC system is comparable to
the step-jump technological achievements of the past in going from
bulk electronics components to miniaturized integrated circuits.
Assuming all of the current design fabrication and technical problems can be resolved in the next few years, integrated optical circuit systems will probably not be available for common consumer applications until late in the 1980's.

In integrated optical circuit high data rate communications
systems, single-mode fibers are the most compatible fiber type since
multimode fibers have very poor coupling efficiencies to single-mode
waveguides. The two common types of coupling methods are the direct
excitation and evanescert field methods. Direct excitation requires
extreme accuracy in mechanical alignments compared to the evanescent
field couplings. For single-mode fibers that have only minute geometric variations on the outer diameter and core concentricity, direct excitation coupling can be accomplished rather efficiently using precision alignment grooves. Evanescent field coupling is used
to couple fiber pairs and thin-film waveguide pairs when they have
similar refractive index profiles. The direct excitation method is
used for coupling thin-film waveguide pairs, fibers to thin-film
waveguides and lasers to fibers when the two components have large
differences in the refractive index profiles.

Several research centers have recently made significant progress in bringing complete integrated fiber optic communication systems closer to a physical reality. For example, IBM has developed
an integrated fiber optic transmitter package mounted on a silicon
wafer containing an array of up to 13 semiconductor lasers fabricated on a single bar of GaAs-GaAlAs. Each laser in the array has
an output power of up to 50 mW and is coupled to the transmitting
fibers with a cylindrical lens with a coupling efficiency of 70%.

References for Chapter 2

Couplers and Coupling

1. Abram, R.A. et al., "The Coupling of Light Emitting Diodes to Optical Fibers Using Spherical Lenses," Journal of Applied Physics, Vol. 46, 1975.

2. Altman, D.E. et al., "An Eight-Terminal Fiber Optics Data Bus Using Tee Couplers," Fiber and Integrated Optics, Vol. 1, 1977.

3. Aoyagi, T. et al., "High-Efficiency Blazed Grating Couplers," Applied Physics Letters, Vol. 29, 1976.

4. Aura, F. et al., "Planar Branching Networks for Multimode and Manmade Glass Fiber Systems," Topical Meeting on Integrated Optics, January, 1976, Salt Lake City, Utah.

5. Auracher, F., "A Photoresist Coupler for Optical Waveguides," Optical Communications, Vol. 12, 1974.

6. Barnoski, M.K., "Data Distribution Using Fiber Optics," Applied Optics, Vol. 14, 1975.

7. Barnoski, M.K. et al., "Angle Selective Fiber Coupler, Applied Optics, Vol. 15, 1976.

8. Bartolini, R.A. et al., "Optical Grating Coupling Between Low Index Fibers and High-Index Film Waveguides," Applied Physics Letters, Vol. 28, 1976.

9. Benson, W.W. et al., "Coupling Efficiency Between GaAlAs Laser to Low Loss Optical Fibers," Applied Optics, Vol. 14, 1975.

10. Blum, F.A. et al., "GaAs Electrooptic Directional Coupler Switch," Applied Physics Letters, Vol. 27, 1975.

11. Boyd, J.T. et al., "Composite Prism-Grating Coupler for Coupling Light into High Refractive Index Thin-Film Waveguides," Applied Optics, Vol. 14, 1976.

12. Brackett, C.A., "On the Efficiency of Coupling Light from Stripe Geometry GaAs Laser into Multimode Fibers," Journal of Applied Physics, Vol. 45, 1974.

13. Burnham, R.D. et al., "Analysis of Grating Coupled Radiation in GaAs: GaAlAs Lasers and Waveguides," IEEE Journal of

Quantum Electronics, Vol.12, 1976.

14. Burnham, R.D. et al., "Grating Coupled Output Beams from Distributed Feedback Diodes Lasers," Proceedings of the Electro-Optics International Laser Conference, 1975, Anaheim, California.

15. Burns, W.K. et al., "Optical Waveguide Parabolic Coupling Horns," Applied Physics Letters, Vol. 30, 1977.

16. Bykovskii, Y.A. et al., "Coupling of Integrated Optics Systems to Fiber Light Guides," Soviet Journal of Quantum Electronics, Vol. 5, 1975.

17. Chang, W.S.C. et al., "Coupling Methods in Prospective Single-Mode Fiber Integrated Optics Systems," Fiber and Integrated Optics, Vol. 1, 1977.

18. Chang, W.S.C., "A New Method for the Efficient Interconnection of High Index Planar Waveguides to Low-Index Transitional Waveguides," Applied Physics Letters, Vol. 29, 1976.

19. Cohen, L.G. et al., "Microlenses for Coupling Injection Laser to Optical Fiber," Applied Optics, Vol. 13, 1974.

20. Fellows, D. et al., "Eccentric Coupler for Optical Fibers: A Simplified Version," Applied Optics, Vol. 14, 1975.

21. Fujita, H. et al., "Optical Fiber Wave Splitting Coupler," Applied Optics, Vol. 15, 1976.

22. Funoyama, T. et al., "Optical Coupler Between Thin-Film Light Waveguide and Optical Fiber," Electronics and Communications in Japan, Vol. 58, No. 2, 1975.

23. Gravel, R.L. et al., "Distributive Tee Couplers," Applied Physics Letters, Vol. 28, 1976.

24. Guttman, J. et al., "Optical Fiber Stripline Coupler," Applied Optics, Vol. 14, 1975.

25. Harris, J.H. et al., "Coupling from Multimode to Single Mode Linear Waveguides Using Horn-Shaped Structures," IEEE Transactions on Microwave Theory and Techniques, MTT-23, 1975.

26. Hsu, H.P. et al., "Single Mode Optic Fiber Pickoff Coupler," Applied Optics, Vol. 15, 1976.

27. Hudson, M.C. et al., "The Star Coupler: A Unique Interconnection Component for Multimode Optical Waveguide Communication Systems," Applied Optics, Vol. 13, 1974.

28. Ishikawa, R. et al., "Micro-Optics Devices for Branching, Coupling, Multiplexing and Demultiplexing," International Conference on Integrated Optics and Optical Fiber Communication, July 1977, Tokyo, Japan.

29. Kamiya, T. et al., "A Large-Tolerant Single-Mode Optical Fiber Coupler with a Tapered Structure," Proceedings of the IEEE, Vol. 64, 1976.

30. Kawaski, B.S. et al., "Optical Directional Coupler Using Tapered Sections in Multimode Fibers," Applied Physics Letters, Vol. 28, 1976.

31. King, F.D. et al., "The Integral Lens Coupled LED," Journal of Electronic Materials, Vol. 4, 1975.

32. Kogelnik, H. et al., "Electrooptically Switched Coupler with Stepped ΔB Reversal Using Ti Diffused $LiNbO_3$ Waveguides," Applied Physics Letters, Vol. 28, 1976.

33. Kohangah, Y., "Injection Laser Coupling to Optical Waveguides with Integral Lenses," Journal of Applied Physics, Vol. 47, 1976.

34. Laybourn, P.J.R. et al., "Optical Coupling Between Thin Films and Circular Fibers," Electronics Letters, Vol. 11, 1975.

35. Lee, A.B. et al., "Optical Access Couplers and a Comparison of Multiterminal Fiber Communications Systems," Applied Optics, Vol. 15, 1976.

36. Lit, J.W.Y. et al., "Fiber-Film Coupling in Integrated Optics," Applied Optics, Vol. 14, 1975.

37. Martin, R.J. et al., "Radiation Fields of a Tapered Film and a Novel Film-to-Fiber Coupler," IEEE Transactions of Microwave Theory and Techniques, MTT-23, 1975.

38. McMahon, D.H., "Efficiency Limitations Imposed By Thermodynamics on Optical Coupling in Fiber Optic Data Links," Journal of the Optical Society of America, Vol. 65, 1975.

39. Nelson, A.R., "Coupling Optical Waveguides By Tapers," Applied Optics, Vol. 14, 1975.

40. Smith, R.B., "Coupling Efficiency of the Tapered Coupler," Electronics Letters, Vol. 11, 1975.

41. Tamir, T., "Beam and Waveguide Couplers," Topics in Applied Physics-Integrated Optics, Vol. 7, 1975.

42. Teh, G.A. et al., "Tapered Optical Directional Coupler," IEEE Transactions on Microwave Theory and Techniques, MTT 23, 1975.

43. Weidel, E., "New Coupling Method for GaAs Laser Fiber Coupling," Electronics Letters, Vol. 11, 1975.

44. Yanai, H. et al., "Tapered Optical Distributed Couplers," Electronics Communication Engineers of Japan, Vol. 75, 1975.

Integrated Optics

45. Albanese, A., et al., "Birefringent Coupler for Integrated Optics," Applied Optics, Vol. 14, 1976.

46. Auracher, F. et al., "Directional Couplers for Integrated Optics," Topical Meeting on Integrated Optics, January, 1976, Salt Lake City, Utah.

47. Ballman, A.A., "The Growth of LiNbO$_3$ Thin Films By Liquid Phase Epitoxy," Ibid.

48. Bartolini, R.C. et al., "Grating Coupling Between Low Index Film and High Index Film Waveguides," Ibid.

49. Blum, F.A. et al., "Monolithic GaAs Circuit Elements for Integrated Optics," Ibid.

50. Boyd, J.T. et al., "Integrated Optical Silicon Photodiode Array," Applied Optics, Vol. 14, No. 6, 1976.

51. Brandt, G.B. et al., "Move on Integrated Optics," Optical Spectra, March 1976.

52. Burns, W.K. et al., "Power Transfer Between Local Normal Modes in Dielectric Waveguide Devices," Topical Meeting on Integrated Optics, January, 1976, Salt Lake City, Utah.

53. Burns, W.K. et al., "Tapered Velocity Couplers for Integrated Optics: Design," Applied Optics, Vol. 14, 1975.

54. Castera, J.P. et al., "Magnetic Stripe Domain Deflector in Integrated Optics," Applied Optics, Vol. 14, 1976.

55. Chinn, S.R., "CdTe Waveguide Devices and HgCdTe Epitaxial Layers for Integrated Optics," Topical Meeting on Integrated Optics, January, 1976, Salt Lake City, Utah.

56. Cohen, L.G. et al., "Microlenses for Coupling Junction Lasers to Optical Fibers," Applied Optics, Vol. 13, No. 1, 1974.

57. Evtuhov, V. et al., "GaAs and GaAlAs Devices for Integrated Optics," IEEE Transactions on Microwave Theory and Techniques, MTT-23, 1975.

58. Giallorenz, T.G. et al., "Performance Limitations I'posed on Integrated Optical Devices By Polarization," Topical Meeting on Integrated Optics, January, 1976, Salt Lake City, Utah.

59. Hunsperger, R.G. et al., "Parallel End-Butt Coupling of a GaAs Laser Diode and a Thin Film Waveguide," Ibid.

60. Kiselyov, V.A. et al., "Determination of Characteristics of Diffused Optical Waveguides," Ibid.

61. Lit, J.W.Y. et al., "Some Uses of Tapers in Integrated Optics," International Conference on Integrated Optics and Optical Fiber Communication, July, 1977, Tokyo, Japan.

62. Logan, R.A., "Integrated Optical Circuits Grown by Liquid Phase Epitaxy," Topical Meeting on Integrated Optics, January, 1976, Salt Lake City, Utah.

63. Mitchell, G.L., "Semiconductor Laser to Integrated Optics Coupling Using Transition Waveguides," Ibid.

64. Sheem, S.K. et al., "Applications of Curved Surface Waveguides in Integrated Optics," Ibid.

Splicing and Connectors

65. Adams, M.J. et al., "Splicing Tolerances in Graded-Index Fibers," Applied Physics Letters, Vol. 28, No. 9, 1976.

66. Bisbee, D.L. et al., "Preparation of Optical Fiber Ends for Low-Loss Tape Splices," Bell System Technical Journal, Vol. 54, 1975.

67. Bisbee, D.L., "Splicing Silica Fibers with an Electric Arc," Applied Optics, Vol. 14, No. 3, 1976.

68. Bisbee, D.L. et al., "A Molded Plastic Technique for Connecting and Splicing Optical Fiber Tapes and Cables," Bell System Technical Journal, Vol. 54, No. 6, 1975.

69. Cherin, A.H. et al., "An Injection-Molded Plastic Connector for Splicing Optical Cables," Bell System Technical Journal, Vol. 55, No. 8, 1976.

70. Curtis, L. et al., "Precision Transfer Molded Single Fiber Optic Connector," Topical Meeting on Optical Fiber Transmission, February, 1977, Williamsburg, Virginia.

71. Dabby, F.W., "Permanent Multiple Splices of Fused-Silica Fibers," Bell System Technical Journal, Vol. 34, No. 2, 1975.

72. Dakss, M.L. et al., "Compensating Fiber Splice Technique," Electronics Letters, Vol. 13, No. 9, 1977.

73. Egashira, K. et al., "Optical Fiber Splicing with a Low-Power CO_2 Laser," Applied Optics, Vol. 16, No. 6, 1977.

74. Fujita, H. et al., "Optical Fiber Splicing Technique with a CO_2 Laser," Applied Optics, Vol. 14, 1976.

75. Gordon, K.S. et al., "Dependency of Dry-Splice Efficiency on Fiber-Break Angle," International Conference on Integrated Optics and Optical Fiber Communication, July, 1977, Tokyo, Japan.

76. Guttman, J., "Multiple Optical Fiber Connector," Electronics Letters, Vol. 11, 1975.

77. Hatakayma, I., et al., "Fusion Splices for Single-Mode Optical Fibers," Topical Meeting on Optical Fiber Transmission, February, 1977, Williamsburg, Virginia.

78. Ikeda, M. et al., "Leaky Modes Effect in Spliced Graded Index Fibers," Applied Physics Letters, Vol. 30, No. 5, 1977.

79. Kimura, T. et al., "Recent Advances in Optical Fiber Transmission Technology," Japan Telecommunications Review, 1976.

80. Kohanzadeh, Y., "Hot Splices of Optical Waveguide Fibers," Applied Optics, Vol. 15, No. 3, 1976.

81. Miller, C.M., "Loose Tube Splices for Optical Fibers," <u>Bell</u>
     <u>System Technical Journal</u>, Vol. 54, 1975.

82. Mochida, K. et al., "Splices for Optical Fibers," <u>Electronics</u>
     <u>Communication Engineers of Japan</u>, Vol. 15, 1975.

83. Smith, P.W. et al., "A Molded-Plastic Technique for Connecting
     and Splicing Optical Fiber Tapes and Cables," <u>Bell System</u>
     <u>Technical Journal</u>, Vol. 54, 1975.

84. Tsuchiya, H. et al., "Loss of Double Eccentric Connectors for
     Optical Fibers," <u>Electronics Communications Engineers of</u>
     <u>Japan</u>, Vol. 75, 1975.

85. Ueno, Y. et al., "A Study of Optical Fiber Connection," Ibid.

CHAPTER 3

LIGHT SOURCES AND MODULATORS

I. Introduction

Light sources for fiber optic communications systems require
certain characteristics including long life-time-in-use, high efficiency, reasonably low cost, sufficient power output, capability
for various types of modulation and physical compatibility with fiber ends. The three light sources that meet these requirements are
semiconductor light emitting diodes (LED's), solid state lasers and
semiconductor injection lasers. A variety of LED's are currently
available for optical signal transmission and an increasing number
of semiconductor lasers.

The most commonly used semiconductor material is the ternary
GaAlAs system since appropriate doping and construction variations
can extend the emission wavelength range to 750 to 1065 nm and the
expected life-time to 5×10^4 hours. Lasers can launch more power
in general, into low NA fibers than LED's. The power output from
lasers however is rather temperature sensitive compared to LED's.
The output temperature-dependence can be controlled with lasers
using special feedback or electronic drive circuits. The cost of
typical DH stripe lasers is about $250 and equipped with tempera-
ture control circuits about $290 in 1977 currency. High-brightness,
small area LED's cost about $160 and low-brightness large area
LED's about $25 each in 1977 currency. By 1981, due to high vol-
ume production and manufacturing, as well as competition of light
sources, the cost of LED's should drop to about $6 and lasers to
about $25 each. As the fiber optics communications field progresses,

94

individual components will become replaced by completed light source modules which will include pre-package miniaturized circuits for temperature stabilization (particularly for lasers and electro-optic modulators) and circuits for output linearization. By 1981 the expected costs of such pre-fab units will be about $35 for lasers and $15 for LED's with expected life-times of about 5×10^4 hours or more.

II. LED's

Light emitting diodes generate light when carriers injected across a p-n junction radiatively recombine, emitting radiation over a solid angle of 2π steradians. The actual output power emitted into a unit solid angle per unit emission area is low for LED's even though the total power output is reasonably high. The low output radiance results in high coupling losses to optical fibers. Thus, most design and manufacturing efforts in LED's for communications systems are aimed at the development of high radiance devices.

The spectral bandwidth of LED's range from 250-400°A at room temperature operation. LED's have a few advantages over lasers in that they are cheaper and have less of a temperature dependence on total emitted power. However, the bandwidth available for communications is smaller. The three basic LED configurations are the edge, high-radiance and surface emitters. The edge emitting hetero-junction structures use partial internal waveguiding to make the emitted light more directional. The lateral width of the emitting region can be adjusted to some extent to match the core diameter of the optical fiber.

A common high radiance LED is the Burrus type, which is a small-area high-brightness DH LED with a brightness up to about $150W/sr-cm^2$ for current densities of $4-6KA/cm^2$ with quantum efficiencies of about 2-3%. This particular LED structure has efficient coupling into the optical fiber. Using an array of microlenses at the bottom of the etched well, it is possible to couple 300 to 400μW

into a fiber with a NA of 0.18.   In coupling light into fiber bundles, the two choices of LED sources are edge emitters and domed
surface emitters.   Domed emitters are rather complex to manufacture
and are more expensive, but have a higher efficiency.

Stripe geometry edge-emitter LED's have the advantage of a
lower junction capacitance compared to the Burrus type LED's, which
leads to a bandwidth of about 100-150MHz for stripe geometries versus 35-50MHz for Burrus LED's.   Although the emission patterns from
Burrus LED's are more easily matched to the acceptance core of the
fiber, the spectral width is wider than stripe geometry LED's.   The
various methods of making stripes include channeled substrates, ox-
ide growth, mesa growth, ion bombardment (by protons or oxygen) and
diffusion methods.

The noise introduced by LED's into optical fibers in most
cases is rather small and is proportional to the information bandwidth and universely proportional to the spectral width and the
number of transmission modes propagated in the fiber.   Modulation
of LED's is accomplished by varying the injected device current.
The limitations of the speed of modulation is dependent upon the
device response time determined electronically by the free carrier
generation and recombination time and the driving circuit capacitance.   The typical power-versus-drive response curves for LED's
are usually linear as shown in Figure 18.   LED's have been the
first light sources used in communications systems since they are
readily available in large quantities at low cost and have large
expected life-time-in-use values up to about 5×10^5 hours.

Two methods commonly used to prepare semiconductor materials
are vapor and liquid phase epitaxy.   Vapor-phase epitaxy is often
used when there is a reasonably large lattice mismatch between the
substrate and the epitaxial layers since the composition can be
graded to any desired degree thus effectively reducing the defect
density in the active region of the grown layer.   Liquid-phase
epitaxy is used especially for the application of Ge, Si and Al
into the semiconductor material layers.   All dopants and materials

FIGURE 18

TYPICAL CURRENT VS. POWER CURVES

FOR OPTICAL LIGHT SOURCES

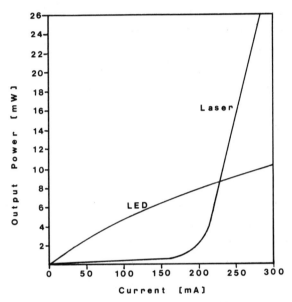

Laser: DH, Stripe Width 20 μm,
Quantum Efficiency 30%
LED: Homojunction, Dome Emitter,
0.4 mm Diameter Surface

used in the construction of LED's and lasers should be of the
highest purity possible to prevent the formation of nonradiative
recombination sites caused by lattice defects and undesirable
chemical precipitates.

III. Lasers

Laser light sources for high data rate transmission systems
are designed to have a high single-pass gain-to-loss ratio to maintain the physical structure as long as possible, thus sustaining
the life-time-in-use at room temperatures. Double-heterojunction
(DH) laser structures can be made with relatively long life-times
when the lattice parameters of the elemental compositions in the
various layers and boundaries are carefully matched. The DH layers
are made by careful epitaxial growth of materials above and below
·the active laser region where light is generated by the radiative
recombination of carriers injected across the p-n junction. The
thin layers surrounding the active region have lower refractive
indices and higher bandgap potentials, thus reducing optical losses
and increasing gain by energy confinement. In the DH laser with
the stripe geometry, the lasing region occurs only in the transverse direction under the width of the contact stripe. Figure 19
shows the typical layer structure of a DH laser.

The development of injection lasers has involved the examination of many materials. The basic semiconductor material for
both LED's and injection lasers is the GaAs combination with a
characteristic emission wavelength of 905nm. In optical communications fibers, one of the OH$^-$ absorption bands covers this wavelength requiring the wavelength to be shifted or the elimination of
the absorption band to achieve the most efficient utilization of
the light energy for low attenuation of signals. The means of
achieving the wavelength shifting is to dope the basic GaAs material with other elements. The elements tested for these purposes
must meet numerous physical and electrical criteria as well as
simple limitation criteria such as having sufficiently high vapor

FIGURE 19

DOUBLE HETEROJUNCTION

INJECTION LASER DIODE

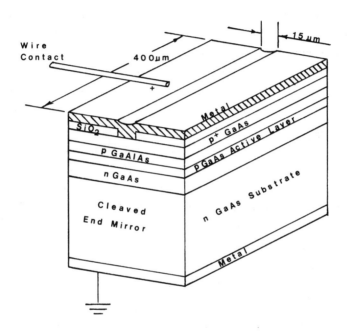

pressures at room temperature operation to maintain material integrity. The elements commonly used in doping of the GaAs material
are Al, As, In, P and Sb, which must be available in ultra-purified
forms to eliminate other metals from generating lattice imperfections in the epitaxial growth stages. The range of experimental
wavelengths achieved by various doping techniques and elemental combinations ranges from 800 to 1150nm.

As in LED's, lasers can be modulated by varying the injection current with modulation bandwidths of several hundred mega-
hertz, or data rates of several hundred megabits per second. Compared to LED's, injection lasers have a high differential quantum
efficiency, in the sense of emitting substantial optical power for
small changes in the driving current above the device threshold
current. The threshold current for typical DH lasers ranges from
0.7 to 14KA/cm^2 and the quantum efficiency from 10 to 50%, compared
to 3% for LED's. The lasing peak energy of heterojunction lasers
decreases with increasing temperature at a rate of approximately
5 x 10^{-4}eV/°K. The threshold current density increases with temperature. The typical power levels needed for optical communications systems is 4 to 12mW. The typical thermal resistance of
stripe lasers is about 11-21°K/W, resulting in a temperature differential between the heat sink and junction of about 5-10°K.

The light output power (mW) versus injection current (mA)
characteristics of stripe lasers depends upon the width of the
stripe region among other factors. For stripe widths less than
10μm the response curves are rather non-linear due to intensity dependent losses. For wider stripes light intensity or current have
little effect on the optical distribution of a given mode. For
stripe widths greater than 30μm the typical output power versus
injection current responses are very linear. Typical chip sizes
use an average stripe width of 15 to 20μm to keep the response
curves reasonably linear by matching the width of the transverse
optical distributions and gain regions. In all semiconductor lasers
some noise is generated caused by quantum fluctuations. This noise

or self-pulsing is greatest near the threshold current level. Wider stripes usually generate less noise except when higher order modes are generated at high current levels. The noise level generated by semiconductor lasers depends mostly upon the drive current and the quality of the laser construction. Below the threshold current the noise level is low, comparable to LED's. The noise level increases rapidly at the threshold current and then decreases at higher current levels. As the current increases sufficiently beyond the threshold, additional lasing modes are usually created causing potentially higher noise levels. Noise is often generated at high frequencies (self-oscillation noise) above 650MHz, which can be effectively eliminated or minimized by limiting the electronic bandwidth of the driving current or by using feedback on the transmitted optical signal.

Table VII lists some commercially available light sources with the drive current, emission wavelength, spectral width and power output. Table VIII lists some of the companies who manufacture or supply LED and laser light sources.

IV. Lifetime

Both LED's and injection lasers have finite life-time's in-use. For LED's the life-times are typically 10^5 hours. For properly designed and prepared lasers, life-times of. 10^5 hours can also be achieved. The two types of failures associated with these communications light sources are gradual degradation and catastrophic degradation. Gradual degradation depends mostly upon the current density and duty cycle. Catastrophic degradation can be caused by many factors including sudden system overload causing surges in the current and physical degradations such as strain releases in the active lasing region. A common cause of catastrophic failure is to overheat a laser diode that does not have temperature feedback current control. The degradation process depends in part on the construction techniques and material purity used in the light source, and the initial concentration of flaws in the active region.

Table VII

Some Typical Commercial Light Sources

LED'S

Manufacturer	Device	Drive Current(mA)	Emission λ_0(nm)	Spectral Width(A)	Power (mW)
ASEA HAFO	1A83	100	940	400	10
Bell-Northern	40-3-30-L	300	840	400	66[a]
Fairchild	FPE500	250	890	400	1[a]
ITT	T851-S	200	840	380	1.5
Laser D. Labs	IRE150	100	820	350	1.5
Meret	TL-36C	300	905	380	12
Monsanto	ME60	50	900	400	0.4
Philips	CQY58	50	875	450	0.5
Plessey	HR954F	300	900	300	35[a]
RCA	SG1009	100	940	450	3.5
Spectronics	SPX2231	100	905	240	2[a]
TI	TIXL472	50	910	230	1

Lasers

Manufacturer	Device	Drive Current(mA)	Emission λ_0(nm)	Spectral Width(A)	Power (mW)
AEG Telefunken	CQX-20	400	820	25	5
ITT	T901-L	350	840	20	7.5
Laser D. Labs	LCW10	200[b]	850	20	14
RCA	C30130	250[b]	820	20	6

a = power units of $W/sr\text{-}cm^2$

b = threshold current in mA

Source: Manufacturers' product sales literature, 1977.
N.B. For further and latest data, contact manufacturers.

Table VIII

Some Manufacturers and Suppliers
of LED or Laser Light Sources

AEG-Telefunken	Monsanto
ASEA-HAFO	National Semiconductor
Bell-Northern	NEC
Fairchild	Philips
Fujitsu	Plessey
Galileo	RCA
General Optronics	Spectronics
Hewlett Packard	Tektronix
Hitachi	Texas Instruments
ITT	Thomson & CSF
Laser Diode Laboratories	Times Fiber Communications
Meret	Valtec

Source: Manufacturers' product sales literature, 1977.
N.B. For further and latest data, contact manufacturers.

Lasers often exhibit two types of rapid decay mechanisms; facet damage and dark-line defects. Facet damage usually results from using an excessive energy density at the laser emitting surface which can be minimized by using small operating power levels and by using dielectric facet coatings such as Al_2O_3. The minimal operational power level is determined by many factors, particularly fiber attenuation and the total length desired or required between end-devices or repeaters. Dark-line defects are caused by the migration of lattice vacancies in the active region which can be minimized by matching the lattices of the surrounding regions and by minimizing externally applied stresses to the laser material. The common cause of externally applied stress to the laser material is

soldering the metal heat sink or contact wire to the top conducting
layer. By using indium solder, a soft low melting temperature solder, residual strain introduced to the laser chip by post-production
heating can be minimized. Internal stresses can be minimized by
adding Al to the active region or P to the passive layers to par-
tially compensate for the coefficient of expansion differences between the $Ga_xAl_{1-x}As$ and GaAs layers.

The conditions normally used for LED's to accelerate life-
time in-use testing is to increase the operational temperature and
extrapolate to a projected failure time or degradation rate for
room temperature operation, which assumes a uniform degradation
process throughout the life-time of the LED. For lasers this means
of testing cannot be used since the operating current cannot be
maintained constant, due to the temperature dependence and time
dependence of the threshold current. The tests usually used to determine life-time utilize rate measurements of the threshold current increase with operational time. These rates are typically
25% threshold current increase per 10 to 15 thousand hours for
lasers with uncoated mirror facets. Coating the facets with dielectric films can effectively eliminate typical current degradation characteristics of the light output with time.

There has been some controversy lately concerning the life-
time ratings of various LED's and laser light sources for fiber
optic applications. Some manufacturers are claiming light source
life-times of 10^5 hours and some up to 10^6 hours. The main problem with all light source life-time ratings is that there is no
standard test for determining the expected life-time in-use. As a
result each manufacturer has developed their own in-house test using
some accelerated time-temperature measurements to determine some
average expected value. Some of these tests are more strenuous or
demanding than others. Before obtaining any light source, laser or
LED, the buyer should inquire what test procedure was used to determine the claimed life-time in-use value, especially in applications
where long component life-times are a necessity (such as underwater
cable repeater components).

Eventually some standard tests or component rating service will be established to alleviate all of the problems centering about manufacturers' claims and ratings for light sources, detectors and other components. In the meantime the consumer should be cautioned that short of actual in-use testing (which requires years) no one really precisely knows what the expected life-time in-use ratings are for commercially available sources. When a standard rating scheme is devised it will be on a relative basis, comparing the tested source to some previously tested standard and not a true absolute value claim for the true life-time in-use under particular environmental and usage conditions.

V. Laser Design

The requirements for communications lasers include output power linearity, low temperature sensitivity, low current thresholds, small lasing areas and long life-times. The causes of non-linearity behavior in the power output versus current responses in lasers is similar to that in LED's. Some of these causes are thermal non-linearity, junction capacitance and uniformity and undesirable radiationless carrier recombinations. Thermal non-linearities can be controlled by using a feedback circuit with a photodetector that monitors the output power from the second laser face. Feedback circuitry also eliminates internal high-frequency modulations.

The design and construction of lasers with emission in the 1000nm region must contend with the following physical and electrical requirements. First, there must be a close matching of the lattice parameters at the heterojunctions to minimize the expected interfacial recombination degradation effects. Second, each layer must have a minimum defect density and maintain low electrical and thermal resistance (10-20°C/W). Third, to keep threshold current densities below 3kA/cm^2 the planar recombination regions must be kept in the range of .1 to .5μm wide and the band gap energy step at the heterojunctions should be at least 0.2eV. Fourth, the p

layer should be doped high enough to prevent high temperature
electron leakage but not so high as to incur excessive current con-
duction away from the lasing area.

The degree of the interfacial lattice parameter mismatch,
(defined as the lattice parameter difference between the substrate
and the epitaxial material divided by the substrate lattice parameter), must be kept small if quantum efficiency is to remain high.
For quantum efficiencies greater than 40%, the lattice parameter
mismatch should not exceed .15% for typical DH lasers. This is
difficult to achieve for any binary material system, although
quantum efficiencies higher than 40% can be attained. The matching
of lattice parameters can be more precisely controlled when multicomponent materials are used in adjoining heterojunction layers.

Many combinations of III-V compounds have been examined as
potential emitters in the 1000nm wavelength region, such as the
InGaAsP alloy which can be matched to the InP compound using liquid
phase epitaxy to achieve fully lattice matched lasers. One criterion used to select the semiconductor materials and elemental
combinations in LED's and laser diodes is the association of band-
structure and radiative lifetime. In indirect bandgap semiconductors such as Si, Ge, GaP and AlAs, the recombination process is
relatively slow. In direct bandgap materials such as GaAs, InAs
and InP the recombination process is very efficient resulting in
short radiative lifetimes. The difference in the radiative life-
times between direct and indirect bandgap materials is about four
to five orders of magnitude. High internal quantum efficiency can
thus be more easily accomplished in direct bandgap semiconductors.

The threshold current density can be minimized by restricting
the width of the recombination region by placing a barrier at a
distance less than the diffusion length from the p-n junction. The
degree of radiation confinement is determined in part by the refractive index differential between the heterojunction layers and
the actual spacing of the layers. The refractive index difference
between GaAs and $Ga_{1-x}Al_xAs$ varies as $0.45X$. The emission wavelength decreases approximately by 6700 X Å $(0 \leq X \leq 0.2)$. In most

lasers, the refractive index step changes at each heterojunction are symmetrical around the active region to optimize wave confinement. The degree of confinement affects the threshold current density and the radiation pattern by changing the effective source size. The threshold current density decreases approximately linearly with the heterojunction layer spacing depth.

Table IX compares laser and LED light sources for a number of emission source parameters.

VI. Non-Semiconductor Lasers

The rare-earth doped solid-state lasers using neodynium (Nd) have potential for fiber optic communications systems because their emission wavelength, ~1065nm, occurs in the low-loss region of glass and silica fibers, and the output beam has a narrow angular distribution. An additional advantage for high capacity transmission systems is that they can generate single-mode outputs with low source noise. The disadvantages of these lasers however are rather significant, some of which are that the lasers are relatively bulky compared to DH lasers and LED's, the life-time in-use is less than other coherent sources and they require external modulation and pumping with resultant low system efficiencies. In addition, many common photodetectors have poor response characteristics at wavelengths around 1065nm, although this problem could be eliminated by altering the detector design and construction.

Non-semiconductor lasers will probably take some time to come to market for some of the same reasons that integrated optical components will take a while to mass-produce, mainly because of the lack óf incentive for very high capacity single-mode fiber optical communications systems. Given the inherent problems of non-semiconductor lasers for optical fiber transmissions, they probably won't emerge as viable alternatives for light sources until further work can mass-produce pre-fab units including all external modulating, pumping and linearization circuitry at reasonable cost. This may not occur until beyond 1985.

Table IX

Comparison of Light Source Types

Light Source Types	Output Power Range(mw)	Bandwidth Range (Mbps)	Spectral Width (Angst.)	Lifetime Average (hours)	Emitter's Area Range(mm^2)	Axial Radiance (W/sr-cm^2)	Source Rise Time(ns)
large area LED	1-7	10-40	350	6×10^4	1-10	0.1-0.5	10-100
small area LED	0.5-1.5	20-200	330	5×10^4	0.004-0.040	5-70	3-20
CW laser diode	3-40	30-900	20	4×10^4	0.002-0.005	10-200	0.1-2
pulsed laser diode	100-400	0.1	20	3×10^4	0.002-0.005	10-200	0.1-2

VII. Modulation

Light sources can be modulated by externally modifying the emitted light after it leaves the source or by directly affecting the source usually by current variations. The typical type of modulation of LED's and semiconductor lasers is by directly modulating the input current in one of three ways. First is to simply use direct modulation which has an intrinsic time delay associated with the rise-time to get above the threshold current. The second way is to bias the laser just below the threshold and modulate the desired signal to levels above the threshold. This eliminates the time delay but adds constant threshold level noise to the outgoing signal. The third method is to bias the laser above the threshold. This third method also eliminates the time delay and threshold noise but adds continual noise (background) to the photodetector, and decreases the expected laser life-time by requiring higher total energy outputs from the laser.

For non-semiconductor lasers such as Nd:YAG lasers, external modulation is used due to pumping, rise-time and fluorescence problems inherent with these solid state devices. External modulators include electro-optic, acousto-optic and magneto-optic reactive modulators and absorptive modulators. Reactive modulators alter the optical signal by changes in electric field, intensity, frequency or phase of the light signal. Magneto-optic modulators have rather significant signal attenuations and must be cooled below room temperature to attain reasonable modulation efficiencies.

Electro-optic modulators work by altering material birefringence in the presence of an electric field, thus affecting the light signal polarization. These modulators usually use intensity modulation of the light source in the transverse mode, with the modulation material (crystal) located between crossed polarizers. The design of electro-optic modulators must take into account possible piezoelectric resonance effects and temperature compensation circuitry.

Acousto-optic modulators work by altering the photoelectric properties of materials to interact with the optical signal, thus creating diffraction and index patterns within the modulator. These modulators have the advantage of having large extinction ratios >30dB limited mostly by light scattering, and require lower drive voltages and powers with better linearity than electro-optic modulators. Although acousto-optic modulators are small, less sensitive to temperature changes and require less power than electro-optic modulators, they are limited in response frequency by the transit time of an acoustic wave across the light beam.

For low modulation frequency requirements, <40MHz, acousto-optic modulators are usually used because of their high efficiency (>95%) at low frequencies. Electro-optic modulators have low optical efficiencies, <30% due to polarizer losses, scattering and thermal effects and are thus used for high modulation frequency applications, >100MHZ. The efficiency of electro-optic modulators decreases with increasing wavelength and with increasing material refractive index values.

A comparison of electro-optic and acousto-optic modulators is given in Table X.

VIII. Source Encoding

Light source encoding alters the information transmission by various techniques to give more efficient transmission per channel. In digital systems minimizing the source bit rate can increase the energy available per transmitted symbol thus lowering the channel BER for a given level of available optical power. In analog transmission source encoding is not used and the signal is modulated by various techniques, including AM, FM, IM, PL and PM. IM analog transmission is the most compatible with existing light sources. Pulse position modulation (PPM) may be used for low bandwidth systems when low dispersion optical fibers are used with light sources capable of high peak power levels at high repetition rates.

Table X

Fiber Optic Modulator Characteristics

Type	Material	λ (nm)	Max.Beam Dia.(mm)	Bandwidth At 650 nm	Insert.Loss At 650 nm
EO	$LiNbO_3$, $LiTaO_3$	600-3500	15	150 MHz	5%
	KDP, ADP	220-1650	45	50 MHz	10%
AO	dense glass	400-2200	10	10 MHz	25%
	$PbMoO_4$	400-2200	10	50 MHz	30%

EO = electro-optic modulator

AO = acousto-optic modulator

Analog transmission sources can be digitized for use in optical systems usually by pulse code modulation (PCM) techniques. Digital transmission is preferred over analog for several reasons at the present stage of development of optical communications. First, signal regeneration at each repeater can be accomplished with a minimal amount of generated noise. Second, analog signals are more susceptible to noise and signal distortion. Third, the frequency division multiplex equipment used for analog systems is more expensive than the time division multiplex equipment used for digital transmission. The use of PCM with light source intensity modulation is relatively insensitive to noise since only the absence or presence of pulse energy is detected. This also permits clean pulse generation for reshaping and amplifying signals in the transmission line.

The three types of multiplexing techniques available for optical systems are frequency division multiplexing (FDM), time division multiplexing (TDM) and space division multiplexing (SDM). For
long length transmission systems where analog signals must be transmitted through one or more repeater-amplifiers, the cumulative effects of signal distortions would be too large for FDM. TDM is
commonly used to multiplex different individual digital sources or
digitized analog sources into one channel, and can be adapted for
optical systems. The initial optical phone links have used optical
sources and transmission with non-optical multiplexing equipment.

Differential PCM can be used to reduce the bandwidth requirements for speech transmission in telephone lines and audio
channels in cable television by reducing the required number of bits
used for encoding, with the use of predictive circuitry. The basic
differential PCM types are DM, CVSDM, VSDM and DCDM (DM = delta
modulation, CVS = continuous variable slope, VS = variable slope
and DC = digital controlled). These types differ mainly in the
types of predictor circuits used. Other more sophisticated types
of prediction algorithms may eventually be used to reduce the required voice channel bandwidth such as LPE (linear predictive encoding) where the spectral envelope is predicted and APD (adaptive
predictive encoding) where both the pitch and spectral envelope are
predicted. PCM requires 64Kbps for quality speech production. Some
of these advanced encodings may permit the required bit rate to
drop to below 10Kbps thus allowing six conversations to occur in
the same space-time frame currently required for one in typical
telephone lines.

References for Chapter 3

Light Sources

1. Abe, M. et al., "GaAlAs LED's for High Quality Fiber Optical
    Analog Link," International Conference on Integrated and Op-
    tical Fiber Communication, July, 1977, Tokyo, Japan.

2. Afromwitz, M.A. et al., "Limitations on Stress Compensation
    in $Al_xGa_{1-x}As_{1-y}P_y$-GaAs LPE Layers," Journal of Applied
    Physics, Vol. 45, No. 11, 1974.

3. Albanese, A., "Automatic Bias Control (ABC) for Injection
    Lasers," Topical Meeting on Fiber Optical Transmission, Febru-
    ary, 1977, Williamsburg, Virginia.

4. Anderson, D.B. et al., "Distributed-Feedback Coupling in
    $Ga_{1-x}Al_xAs$ Double-Heterostructure Lasers: Effect of Aluminum
    Concentration," Applied Optics, Vol. 14, No. 9, 1975.

5. Apruzzese, J. et al., "Linear Light-Output Characteristics in
    Double Heterostructure GaAlAs Lasers," International Conference
    on Integrated Optics and Optical Fiber Communications, July,
    1977, Tokyo, Japan.

6. Appert, J.R. et al., "Optical Properties of Vapor-Grown In_x-
    $Ga_{1-x}As$ Epitaxial Films on GaAs and $In_xGa_{1-x}P$ Substrates,"
    Journal of Applied Physics, Vol. 45, No. 1, 1974.

7. Bellavane, D.W. et al., "Room Temperature Mesa Lasers Grown
    By Selective Liquid Phase Epitaxy," Applied Physics Letters,
    Vol. 29, 1976.

8. Brady, J. et al., "GaAs Laser Source Package for Multi-Channel
    Optical Links," Topical Meeting on Fiber Optical Transmission,
    February, 1977, Williamsburg, Virginia.

9. Burnham, R.D. et al., "Distributed Feedback Buried Heterostruc-
    ture Diode Laser," Applied Physics Letters, Vol. 29, 1976.

10. Burns, C.A. et al., "Nd:YAG Single Crystal Fiber Laser: Room
    Temperature CW Operation Using a Single LED As an End Pump,"
    Ibid.

11. Cantagrel, M. et al., "Reproducibility of the Manufacturing Process of the CW GaAs Lasers," Second European Conference on Optical Fiber Communication, September, 1976, Paris, France.

12. Capik, R.J. et al., "Efficient Lattice-Matched Double Heterostructure LED's at 1.1µm from GaInAsP," Applied Physics Letters, Vol. 28, 1976.

13. Casey, H.C. et al., "Influence of $Al_xGa_{1-x}As$ Layer Thickness on Threshold Current Density and Differential Quantum Effiviency for $GaAsAl_xGa_{1-x}As$ DH Lasers," Journal of Applied Physics, Vol. 46, No. 3, 1975.

14. Chinone, N. et al., "Long-Term Degradation of $GaAs-Ga_{1-x}Al_xAs$ DH Lasers Due to Facet Erosion," Journal of Applied Physics, Vol. 48, No. 3, 1977.

15. Dawson, L.R., "High-Efficiency Graded Band-Gap $Ga_{1-x}Al_xAs$ Light-Emitting Diodes," Journal of Applied Physics, Vol. 48, No. 6, 1977.

16. Dierschke, E.G., "Light Emitting Surfaces for Optical Waveguides," Proceedings of SPIE, Vol. 60, 1975.

17. Dixon, R.W. et al., "Reliability of DH GaAs Lasers at Elevated Temperatures," Applied Physics Letters, Vol. 26, 1975.

18. Eliseev, P.G. et al., "Role of Mechanical Stresses in Gradual Degradation of Light Emitting Diodes and Injection Lasers," Soviet Journal of Quantum Electronics, Vol. 5, 1975.

19. Epstein, M. et al., "Efficient White Laser Illuminators for Plastic Optical Fibers," Applied Optics, Vol. 16, No. 4, 1977.

20. Ettenberg, M. et al., "Low-Threshold Double Heterojunction AlGaAs/GaAs Laser Diodes:Theory and Experiment," Journal of Applied Physics, Vol. 47, 1976.

21. Ettenberg, M. et al., "Vapor Grown CW Room Temperature GaAs/InGaP Lasers," Applied Physics Letters, Vol. 29, 1976.

22. Farmer, G.I. et al., "Low Current Density LED-Pumped Nd:YAG Laser Using a Solid Cylindrical Reflector," Journal of Applied Physics, Vol. 45, No. 3, 1974.

23. Goodwin, A.R. et al., "Threshold Temperature Characteristics of Double-Heterostructure $Ga_{1-x}Al_xAs$ Lasers," Journal of Applied Physics, Vol. 46, No. 7, 1975.

24. Goodwin, A.R. et al., "Semiconductor Lasers for Optical Communications," Second European Conference on Optical Fiber Communication, September, 1976, Paris, France.

25. Groves, W.O. et al., "Multiple Liquid Phase Epitaxy of $In_{1-x}Ga_xP_{1-z}As_z$ Double Heterojunction Lasers: The Problem of Lattice Matching," Applied Physics Letters, Vol. 31, No. 1, 1977.

26. Harth, W. et al., "Time and Frequency Response of Different Types of Light-Emitting Diodes," Second European Conference on Optical Fiber Communication, September, 1976, Paris, France.

27. Hayashi, I., "Semiconductor Light Sources," International Conference on Integrated Optics and Optical Fiber Communication, July, 1977, Tokyo, Japan.

28. Henderson, D.M., "Waveguide Lasers with Intracavity Electro-optic Modulators: Misalignment Loss," Applied Optics, Vol. 15, No. 4, 1976.

29. Hsieh, J.T., "Room Temperature Operation of GaInAsP/InP Double-Heterostructure Diode Lasers Emitting at 1.1μm," Applied Physics Letters, Vol. 28, 1976.

30. Hwang, C.J., "Initial Degradation Mode of Long-Life (GaAl)As-GaAs Lasers," Applied Physics Letters, Vol. 30, No. 3, 1977.

31. Ichikawa, S. et al., "Small Package Optical Semiconductor Devices for Optical Fiber Data Transmission Systems," International Conference on Integrated Optics and Optical Fiber Communication, July, 1977, Tokyo, Japan.

32. Ito, R. et al., "Buried-Heterostructure Lasers as Light Sources in Fiber Optic Communications," Ibid.

33. Kressel, H. et al., "Light Sources," Physics Today, May, 1976.

34. Lee, C.P. et al., "Embedded GaAs-GaAlAs Heterostructure Lasers," Topical Meeting on Integrated Optics, January, 1976, Salt Lake City, Utah.

35. Lee, T.P., "Effect of Junction Capacitance on the Rise Time of LED's and on the Turn-On Delay of Injection Lasers," Bell System Technical Journal, Vol. 54, 1975.

36. Logan, R.A. et al., "GaAs Double Heterostructure Lasers Fabricated by Wet Chemical Etching," Journal of Applied Physics, Vol. 47, 1976.

37. Ludowise, M.J. et al., "Variation of Effective Index of Refraction in a Double-Heterojunction Laser $(In_{1-x}Ga_xP_{1-z}As_z)$ Journal of Applied Physics, Vol. 48, No. 5, 1977.

38. Marcuse, D., "Excitation of Parabolic-Index Fibers with Coherent Sources," Bell System Technical Journal, Vol. 54, 1975.

39. McMullin, P.G. et al., "Effect of Doping on Degradation of GaAsAl-GaAs Injection Lasers," Applied Physics Letters, Vol. 24, 1974.

40. Melngailis, J., "Lasers at 1.0-1.3μm for Optical Fiber Communications," International Conference on Integrated Optics and Optical Fiber Communication, July, 1977, Tokyo, Japan.

41. Mockel, P.G., "Properties of an Epitaxially Grown Nd:YAG Waveguide Laser," Topical Meeting on Integrated Optics, January, 1976, Salt Lake City, Utah.

42. Nagai, H. et al., "In $P/Ga_xIn_{1-x}As_yP_{1-y}$ DH LED's in the 1.5μm Wavelength Region," International Conference on Integrated Optics and Optical Fiber Communication, July, 1977, Tokyo, Japan.

43. Nagai, H. et al., "LPE of $GaAs_xSb_yP_{1-x-y}$ Continuously Graded Composition Layer," Journal of Applied Physics, Vol. 47, No. 12, 1976.

44. Nahory, R.E. et al., "Continuous Operation of 1.0μm Wavelength GaAsSb/AlGaAsSb Double-Heterostructure Injection Lasers at Room Temperature," Applied Physics Letters, Vol. 28, 1976.

45. Nakamura, M., "Distributed Feedback Semiconductor Lasers," Topical Meeting on Integrated Optics, January, 1976, Salt Lake City, Utah.

46. Nuese, C.J. et al., "Room Temperature Heterojunction Laser
     Diodes of the InGaAs/InGaP with Emission Wavelength Between
     0.9 and 1.15μm," Applied Physics Letters, Vol. 26, 1975.
47. Parish, M.B., "Heterostructure Injection Lasers," IEEE Trans-
     actions on Microwave Theory and Techniques, MTT-23, 1975.
48. Seib, D.H. et al., "Multiply-Resonant Distributed Feedback
     Lasers," Topical Meeting on Integrated Optics, January, 1976,
     Salt Lake City, Utah.

Modulators

49. An, J.C. et al., "Temperature-Compensated Optical Modulators
     Using the Feedback Control Method," Electronics and Communica-
     tions in Japan, Vol. 58, No. 8, 1975.
50. Arragon, J.P., "Practical Limits of the Modulation Speed of
     Light-Emitting Diodes," Second European Conference on Fiber
     Optic Communication, September, 1976, Paris, France.
51. Burns, W.K. et al., "Active Branching Waveguide Modulator,"
     Applied Physics Letters, Vol. 29, 1976.
52. Cheo, P.K. et al., "Recent Development on High Power Infrared
     Waveguide Modulators," Topical Meeting on Integrated Optics,"
     January, 1976, Salt Lake City, Utah.
53. Davies, D.E.N. et al., "Method of Phase-Modulating Signals in
     Optical Fibers: Application to Optical-Telemetry Systems,"
     Electronics Letters, Vol. 10, 1974.
54. DeLaRue, R.M. et al., "Electro-Optic Y Junction Modulator/
     Switch," Electronics Letters, Vol. 12, 1976.
55. Gravel, R.L. et al., "Electro-Optic Channel Modulator for
     Multimode Fibers," Applied Physics Letters, Vol. 28, 1976.
56. Hammer, J.M. et al., "Stripe Waveguides and Coupling Modula-
     tors of Lithium Niobate Tantalate," Topical Meeting on Inte-
     grated Optics, January, 1976, Salt Lake City, Utah.
57. Holden, W.S., "Pulse Position Modulation Experiment for Optical
     Fiber Transmission," Topical Meeting on Optical Fiber Trans-
     mission, January, 1975, Williamsburg, Virginia.

58. Itoh, T. et al., "Broad-Band LiNbO$_3$ Optical Waveguide Modulator," International Conference on Integrated Optics and Optical Fiber Communications, July, 1977, Tokyo, Japan.

59. Izutsu, M. et al., "Optical Waveguide Modulator Using Electro-optic Crystal as a Substrate," Electronics and Communications in Japan, Vol. 58, No. 9, 1975.

60. Kaminow, I.P. et al., "Optical Waveguide Modulators," IEEE Transactions on Microwave Theory and Techniques, MTT-23, 1975.

61. Kingsley, S.A., "Optical-Fiber Phase Modulator," Electronics Letters, Vol. 11, No. 19, 1975.

62. Kleinman, D.A. et al., "Vibration-Induced Modulation of Fiber-guide Transmission," Topical Meeting on Optical Fiber Transmission, February, 1977, Williamsburg, Virginia.

63. McKenna, J. et al., "Polarization Modulation with Double Heterostructure Pn Junction Diodes," Topical Meeting on Integrated Optics, January, 1976, Salt Lake City, Utah.

64. Nagura, R. et al., "Pulse Detection Characteristics and Modulation Limit of Optical PCM Communication Using Envelope Modulation," Electronics and Communications in Japan, Vol. 58, No. 4, 1975.

65. Nelson, A.R. et al., "Active Components and System Concepts for Multiterminal Fiber Optic Data Links," Proceeding of Intel Comm, October, 1977, Atlanta, Georgia.

66. Reinhart, F.K. et al., "Phase Modulation Nonlinearity of Double-Heterostructure Pn Junction Diode Light Modulators," Journal of Applied Physics, Vol. 45, No. 5, 1974.

67. Tangonan, G.L., "Development of Thin Film Modulators with High Throughput and High Optical Power Handling Capabilities," International Conference on Integrated Optics and Optical Fiber Communications, July, 1977, Tokyo, Japan.

68. Ueno, Y. et al., "A Semiconductor Laser Communications System Using Pulsed-Internal Modulation," Electronics and Communications in Japan, Vol. 59, No. 4, 1976.

CHAPTER 4

PHOTODETECTORS AND REPEATERS

I. Introduction: Photodetectors

Photodetectors used for fiber optic systems must meet numerous physical and electrical requirements including: adding a minimal amount of noise to the transmission signal, having sufficient
bandwidth (speed of response) to track light intensity variations in
the transmitted light, peak sensitivity at the light source wavelength, stability over changing external temperatures, and the detectors must have long life-times in-use at reasonable cost.

Semiconductor photodiodes all have a depleted region where
electron-hole pairs are generated by the absorbed photons, with a
high electric field between two semiconducting regions. To prevent
the formation of local high-field regions at the edges of the junction (where the electron-hole pairs are collected) guard-rings are
often used in the diode structures. Fiber optic systems usually
involve relatively low light levels, so the detectors are operated
with a reverse bias such that the output current varies very
linearly with the intensity of the incident optical signal.

The probability of light absorption in the depletion region
depends upon the wavelength of the incident signal, the type of
semiconductor material used and the thickness of the depletion region. For each semiconductor material, once a certain critical
wavelength has been exceeded, the material becomes almost transparent to incident radiation. The thickness of the depletion region can be increased to extend the wavelength before the trans-
parency frequency is reached but this extension is greatly limited

119

by the necessary bandwidth (speed of response) of the detector.

Photodiodes are usually described by four basic quantities; response time, quantum efficiency, total noise equivalent power and responsivity. The response time is the transit time for electrons in the photodiode to transverse from the cathode to anode, typically from 0.2 to 5ns. Table XI lists the properties of various commercial PIN photodetectors. The quantum efficiency is the percentage of incident photons that liberate photodetector electrons. The responsivity is the average emitted current divided by the average incident power. Figure 20 shows the quantum efficiency and responsivity versus wavelength for typical PIN photodiodes. In avalanche photodiodes (APD) the two basic entities usually considered are the response speed and the multiplication or gain characteristics. The physical structure of avalanche diodes usually includes a guard ring to prevent excessive leakage at the junction edges and low breakdown voltages. Many APD devices are silicon based with anti-reflection coatings to provide quantum efficiencies near 90% with gains of several hundred. Table XII lists the properties of various commercially available APD devices.

II. Responsivity, Current, and TNEP

The responsivity of silicon decreases rapidly for wavelengths beyond 990nm. Thus, for both PIN and APD devices other elemental systems must be used to attain high-speed detectors in the wavelength region extending beyond 990nm. Typical PIN detectors are usually rated for 200 to 1100nm with the responsivity peaks at 900nm or less. The typical operating voltages range between 10 to 110 volts with active surface areas ranging from 1mm^2 to 10mm^2. Responsivities typically range from 0.2 to 0.7A/W with minimum detectable light levels from 1 to 10 x 10^{-15}W and noise equivalent powers ranging from 1 to 10 x 10^{-14}W/Hz$^{\frac{1}{2}}$. Increasing the responsivity of existing silicon PIN detectors to wavelengths above 990nm usually involves increased bias voltages and subsequent greater sensitivity to temperature changes. In addition, designs for longer wavelengths usually require greater junction capacitances

Table XI

Typical Commercial PIN Photodetectors

PIN Diodes

Manufacturer	Device	Responsivity(A/W) At λ_o (nm)		Response Speed(ns)	Bias Volt.(V)
Bell-Northern	BNRD-5-1	0.55	(840)	1	100
EG&G	FND-100	0.62	(900)	1	90
Galileo	5105-031	0.50	(905)	5	50
Harshaw	538	0.62	(900)	7	100
Hewlett-Pack.	5082-4205	0.45	(900)	1	20
Infrared Ind.	7258	0.51	(1060)	8	50
RCA	C30808	0.65	(900)	5	45
Spectronics	SD5426	0.64	(905)	2	50
TI	TIXL80	0.55	(950)	15	100
UDT	PIN-3D	0.40	(850)	15	50

Preamp Hybrid Modules

		(mA/uW)		3dB Bandwidth
Bell&Howell	529	40	(800)	0.1
Centronic	OSI5LHSB	15	(900)	1
EG&G	MHZ-018Y	40	(900)	40
Meret	MDA425	15	(900)	10
RCA	C30818	300	(850)	40
TI	TIXL151	230	(900)	50

Source: Manufacturers' product sales literature, 1977.
N.B. For further and latest data, contact manufacturers.

FIGURE 20

QUANTUM EFFICIENCY VS. WAVELENGTH FOR

TYPICAL SILICON PIN PHOTODIODES

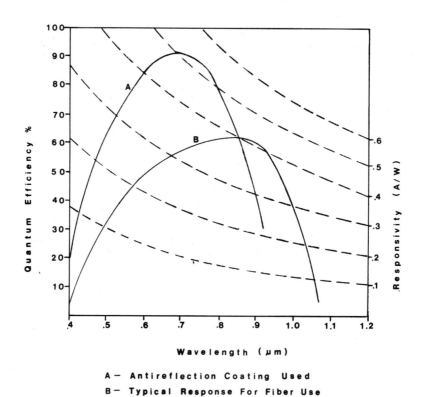

Wavelength (μm)

A — Antireflection Coating Used
B — Typical Response For Fiber Use

Table XII

Typical Commercial APD Devices

Manufacturer	Device	Responsivity(mA/uW) At λ_0 (nm)		Bandwidth (MHz)	Bias Volt.(V)
AEG	BPW28	0.20	(850)	200	170
EMI	S30500	0.04	(880)	200	165
GE	50CHS	0.10	(900)	100	200
RCA	30884	0.63	(900)	250	275
RCA	30817	0.16	(1060)	200	375
TI	TIXL451	0.20	(900)	200	100
TI	TIXL452[a]	75	(900)	40	230

a = fiber optic module amplifier unit

Source: Manufacturers' product sales literature, 1977.
N.B. For further and latest data, contact manufacturers.

resulting in bandwidth reduction. For these reasons researchers and designers are attempting to develop other materials and device structures for use at wavelengths above 990nm. Table XIII compares the typical properties of PIN and APD detectors.

The average current produced by a photodetector is given by:

$$I\,(t) = G\left(\frac{2\pi Q\lambda_0}{hc}\ P\,(t) + I_d\right)$$

where G is the average gain, Q is the quantum efficiency, λ_0 is the nominal source wavelength, h is Plank's constant, P(t) is the optical power incident on the photodetector (time dependent) and I_d is the dark current.

The total noise equivalent power, TNEP, accounts for both the noise level production in the devices and its gain to provide a

Table XIII

Comparison of Detector Types

Basic Detector Type	Sensing Diameter Area (mm^2)	Sensitiv. At 1 MHz (dBm)	Responsiv. At Peak λo (A/W)	Bias Voltage (V)	Dynamic Range (dB)	Maximum Data Rate (Mbps)	Lifetime Range (hours)	Rise Time (ns)	Peak λo Response (nm)
PIN	0.3-3	-58	0.4-0.7	10-100	60	1-2 GHz	10^4-5×10^5	1-5	870
APD	0.8-8	-70	10-70	250-350	20	90-150 MHz	10^4-3×10^5	2-5	880

measure of the minimum detectable signal. The TNEP is defined as
the amount of light energy impinging on the active surface area
which will produce an output signal equal in magnitude to the noise
output of the photodetector device. The units typically used for
TNEP are nW. The NEP of a detector is the noise equivalent power
which is simply the TNEP normalized to a 1Hz bandwidth and is given
in units of W/Hz.

For the value of NEP to be meaningful, the load and amplifier resistances should be specified. The basic circuit diagram
for an APD is shown in Figure 21. The circuit for a PIN is the
same except for the power supply unit which is of lower voltage and
requires less control than the APD. The APD devices all have a
large responsivity dependence on the bias voltage and temperature
as shown in Figure 22. Since the detector breakdown voltage varies
with temperature the precise bias needed to maintain a fixed gain
varies. Precise control is maintained by current feedback control
or by monitoring the gain in an auxiliarry APD matched to the active diode device. Table XIV lists some manufacturers and suppliers of PIN and APD photodiodes.

The formula used to calculate TNEP for PIN detectors is:

$$TNEP_{PIN} = \frac{4hfB}{Qe} (2kTC)^{\frac{1}{2}},$$

where h is Planck's constant, f is the transmission optical frequency, B is the bandwidth, Q is the device quantum efficiency, e
is the electron charge, k is Boltzmann's constant, T is the diode
amplifier temperature and C is the diode capacitance. For a given
optical system, the TNEP can be optimized by reducing the detector
capacitance and increasing the quantum efficiency. Detector capacitance depends mostly on the junction area and the width of the
depletion region. For APD detectors, the formula to calculate the
TNEP is given by:

$$TNEP_{APD} = \frac{hfB}{QeG} (2kTC)^{\frac{1}{2}},$$

FIGURE 21

APD SCHEMATIC DIAGRAM

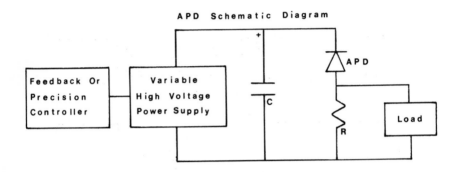

APD Schematic Diagram

where G is the detector gain level where the shot noise and thermal
noise at the temperature T are equal. The gain G, typically
ranges from 10 to 100. The value of G which minimizes the NEP is
not necessarily the optimal value of G to use at high SNR levels.

Although the APD devices have a reduced TNEP, they must
operate at higher voltage and require a greater degree of control
on current and voltage. APD devices with 1GHz bandwidth are available with response times less than 1ns and a TNEP less than 10μW.
The NEP of typical detectors can be determined as a function of the
gain for a specified bandwidth. The degree of gain in a given APD
depends on the ratio of the hole collision ionization probability
to the electron collision ionization probability given by K (not
to be confused with the Boltzman constant). For quality silicon
detectors the value of k is usually about 0.025.

LED's, lasers and photodetectors used in optical fiber communications systems all have the characteristic of changing certain
electrical and signal properties as a function of temperature. For

FIGURE 22

APD RESPONSIVITY VS. VOLTAGE AS

A FUNCTION OF TEMPERATURE

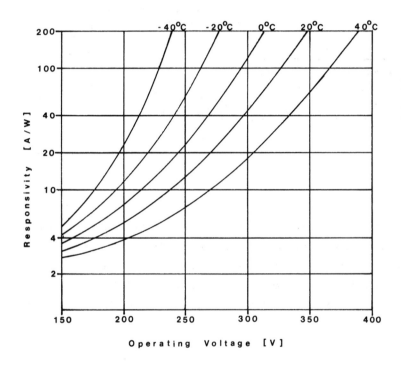

detectors without any temperature compensation circuitry coupled to
the device the wavelength of the peak responsivity is shifted about
0.4 to 0.5nm/°C. The emission wavelength of LED's typically shifts
0.3nm/°C when no temperature feedback circuitry is used. Since the
emission wavelength of the light source and the peak responsivity
of the detector can shift at different thermal rates, it is possible
in some applications to incur significant efficiency losses. For
example if the light source is located indoors and a series of de-
tector-repeaters are located outdoors, the mismatching of transmission wavelength and responsivity can accumulate to a total system efficiency reduction of 8% of the retransmitted signal per
detector-repeater unit. Using carefully designed compensation circuitry to maintain constant transmission wavelength and detector
peak responsivity, is critical to any high data rate system where
operational temperature differentials exist between system components.

III. Detector Coupling

The coupling of detectors to fiber ends presents few of the
problems that coupling light sources to fibers present since angular displacement and lateral displacement do not cause significant
losses in detectors since they have relatively large active areas
compared to the fiber core and because detector surfaces are not
sensitive to the incidence angle of light. The major cause of
coupling losses with detectors is by reflection. For detectors
with flat windows covering the active area reflection losses can be
rather high in the air interface region between fiber end and detector window. Anti-reflection coatings can eliminate part of
these losses or index matching epoxies can be inserted at the interface to improve efficiencies. Placing the detector surface in
direct contact with the fiber end bonded by a clear non-absorbing
epoxy represents one of the lowest insertion loss couplings between
detectors and fibers. For available PIN diodes coupling losses
range from 0.2 to 4dB depending upon the detector casing surface,

Table XIV

Some Manufacturers and Suppliers
of PIN or APD Photodetectors

AEG-Telefunken	Hughes Aircraft
American Electronic Laboratories	Infrared Industries
ASEA-HAFO	ITT
Bell-Northern	Meret
Bell & Howell	Motorola Semiconductor
Centronic	NEC
Devar	Nuclear Equipment Corporation
EG&G	Optoelectronics
Electro-Nuclear Laboratories	Philips
EMI Electronics	Plessey
Fairchild Semiconductor	Quadri
Ferranti	Quantrad
Fort	Raytheon
Fujitsu	RCA
Galileo	Spectronics
General Electric	Texas Instruments
General Instrument	Thomson & CSF
Harshaw	Twentieth Century Electronics
Hewlett Packard	United Detector Technology

the use of antireflection coatings and the coupling technique used.

As with lasers and LED's the linearity of response of all
fiber optic components is important especially for high data rate
analog systems. The response of PIN diodes is extremely linear
even at low bias voltages, and the measurable non-linearities approach the noise level of the device except at extremely low (unus-
able) bias voltages. PIN detectors have rather low responsivity
and slow response speed for wavelengths above 990nm and have no internal gain as compared to APD devices. In choosing a detector for
a particular communications system, some of the parameters that
must be considered are responsivity at the transmission wavelength,
bandwidth, noise-equivalent power, device life-time and cost, and
environmental considerations (temperature, applied stress). All of
these parameters must be considered when attempting to match the
smallest received signal to the smallest detectable signal without
reducing the system capacity, bit error rate (BER) or signal-to-
noise ratio (SNR). APD devices are usually used over PIN's in
situations where optimal noise performance is needed since the
noise characteristics of PIN's are dominated by the preamplifier
thermal noise. Other potential sources of noise are shot noise
and 1/f noise. PIN's are usually used for small bandwidth systems
<15MHz and short-length runs <500m. Improvement on the bandwidth
capabilities can be achieved by matching the active detector area
to the fiber area, which reduces device capacitance.

IV. Detector Modules

Detectors are available as single devices or as a module
unit coupled to a preamplifier. These modules are small, light
weight, reliable, have enhanced frequency response (stray induc-
tances and capacitances eliminated) and have reduced stray noise
(detector-preamp unit is in a shielded environment). A large number of these modules are available commercially ranging in price
from $20 to $200 with bandwidths ranging from 0.1 to 50MHz and
responsivities from 10 to 500mA/W at peak wavelengths from 800 to

950nm. The preamplifiers used in the modules are low impedance operational amplifiers (op-amps) with fast response speeds and large linear dynamic ranges. The important limiting factor in op-amps for fiber optic systems is the noise level or SNR. For many modules the actual usable bandwidth is only about 5 to 10MHz due to the large op-amp noise spectral density. The front-end of the amplifier can be designed to fit the system data rate in the sense that bipolar integrating front-ends have a better SNR for rates >25Mbps while silicon FET's are better for rates <15Mbps.

V. Temperature Control

As with lasers and LED's, PIN performance and linearity is effected by temperature effects which can be compensated for by an appropriate feedback loop signal circuitry. Even the best designed temperature compensated modules have linearities less than that of the individual components which can present problems for large bandwidth, high gain systems. APD's have high internal gains, high responsivity, fast response times and large gain-bandwidth products, with small active areas. As the gain factor increases, the need for precise controls on voltage and temperature become more critical if a constant gain is to be achieved. For most applications the temperature should be controlled to within .1°C and the voltage to within .1V with appropriate feedback circuitry to avoid non-linearities, excess noise and drifting. Typical APD's have responsivity ratings at the spectral peak ranging from 10 to 100 with bias voltages ranging from 100 to 400 volts. The bandwidth rating for APD's ranges from 100 to 500MHz with prices ranging from $150 to $350.

At low light signal levels at the detector end of the fiber the APD is usually used instead of PIN diodes because of the high internal gain characteristics. The usual choice of preamplifiers are trans-impedance types which have a wide dynamic range but not as good as SNR compared to integrating preamplifiers. APD's are available in pre-fabricated modules with preamplifiers and regulated

power supplies for temperature compensation. The modules were expensive in 1977, ranging from $250 to $450, but the cost should come
down by 1981 with mass-production of fiber optic components. The
present bandwidth of these modules ranges from 30 to 300MHz. At
the higher bandwidths, the signal level above the noise is dramatically reduced.

VI. Introduction: Repeaters

The basic optical repeater consists of a photodetector,
amplifier, equalizer and signal regenerator coupled to an optical
driver and light source. Figure 23 shows the basic block diagram
for a typical PCM receiver. The equalizer in the circuit compensates for the effects of the photodetector and the frequency-time
response of the optical fiber line. The objective of the equalization unit in the receiver is to eliminate or reduce the intersymbol
interferences. Equalization imposes a power penalty on the total
receiver unit since the equalizing filter is not optimal in terms
of noise performance. The equalizer response increases with frequency, enhancing the noise level. The types of equalizers that
can be used are LC (passive or transversal), RC (active or transversal) or distributed transversal equalizers. If the communications system is attenuation limited the equalizer is not needed.
In systems limited by dispersion however, the use of the equalizer
can extend repeater spacings. Table XV lists some commercial receivers and their properties. Table XVI lists some manufacturers
and suppliers of repeaters and optic links.

Optical repeaters are very similar to conventional wire system repeaters with the obvious differences of the light detector
and drive-source components in the circuitry. The main difference
in the design and analysis of receivers for optical systems and
conventional wire systems is the receiver noise characteristics
which for optical repeaters can be rather complex since both
Gaussian and Poisson noise components are present, some of which
are dependent upon the transmitted signal. The front end amplifiers

FIGURE 23

BASIC BLOCK DIAGRAM FOR PCM RECEIVER

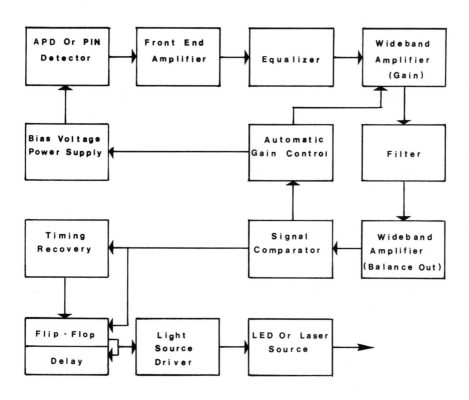

in optical repeaters must be carefully designed to function with
the detectors (capacitive devices), and the optical driver designed
for the light sources (low impedance, high current devices). The
light source driver must be designed to deal with the transient response characteristics of the light source to reproduce the desired
amplified and shaped pulses coming from the rest of the repeater.
This is accomplished by internally correcting for the light source
response or by a feedback mechanism which responds to the output
light signal. The amplifiers used in the repeater are either FET
or bipolar discrete or some hybrid combination.

        Repeaters have been built and tested for various types of
transmission signals (analog and digital) at bit rates ranging
from 10 - 200Mbps. The higher bit rate repeaters (>25Mbps)
use APD detectors with automatic gain control circuitry. The automatic gain control section of the receiver compensates for changes
in the incoming signal level, amplifier gains and the dramatic effects of temperature in the APD device. Analog repeaters are more
complex in design compared to digital repeaters where the information signals can be completely and clearly regenerated in each
successive repeater with little signal deterioration (very small
BER's). In analog transmission each repeater-amplifier adds additional noise to the signal which limits the number of repeaters
that can be used in the transmission line as a function of the input
SNR, the associated repeater noise level and the specification
limit on the final received SNR to the end-device. The minimum required received power (receiver sensitivity) is a function of bit
rate in optical repeaters and varies greatly depending upon the optical component and amplifier designs and the use of detector gain.
Figure 24 shows the minimum required receiver power versus bit rate
for APD and PIN devices using integrating bipolar amplifiers, using
fibers with nominal pulse dispersion.

Table XV

Typical Commercial Receiver-Transmitter Modules

Manufacturer	Device	3 dB Bandwidth	Detector Responsivity (A/W) At λ_o (nm)	
Bell-Northern	BNR-R6	20 MHz	0.50	(840)
Centronic	FRD-1-4	6 MHz	0.36	(900)
Fujitsu	PCM-6M	6 MHz	0.50	(840)
Galileo	3555971R	2 MHz	0.50	(905)
Plessey	OML30D	30 MHz	0.50	(900)

Source: Manufacturers' product sales literature, 1977.
N.B. For further and latest data, contact manufacturers.

Table XVI

Some Manufacturers and Suppliers
of Fiber Optic Links or Repeaters

ASEA-HAFO	Laser Diode Laboratories
Bell-Northern	Meret
Belling & Lee	Optical Communications Corp.
Centronic	Plessey
Fort	Quadri Corporation
Fujitsu	RCA
Furukawa Electric	Siecor
Galileo	Spectronics
Harris	Thomson & CSF
Hellerman-Deutsch	Twentieth Century Electronics
ITT	Valtec

Source: Manufacturers' product sales literature, 1977.
N.B. For further and latest data, contact manufacturers.

FIGURE 24

REQUIRED OPTICAL POWER VS. BIT RATE

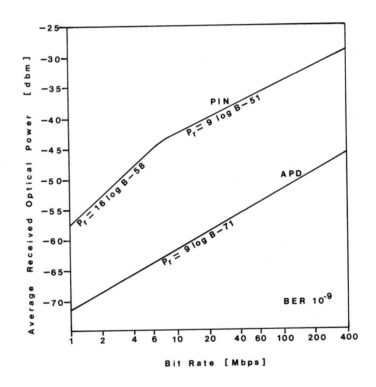

VII. Repeater SNR

For an optical receiver using an APD followed by a baseband amplifier with a noise figure N_f with an equivalent resistance R to the APD, the SNR for a sinusoidal signal is given by:

$$SNR = \frac{(QeGmP_0)^2}{2(hvi_n)^2}$$

where P_0 is the average received optical power, m is the modulation index of the signal (-1 for 100% modulation) and i_n^2 is the mean square noise current after avalanche gain. The mean square noise current is given by:

$$i_n = i_q + i_T + i_d + i_1 + i_g + i_b,$$

the sum of the quantum, thermal, dark, leakage, background and beat noise currents respectively. These currents are defined as follows:

$$i_q^2 = \frac{2e^2}{hv} QP_0 G^2 N_d B$$

$$i_T^2 = \frac{4KT}{R} N_t B$$

$$i_d^2 = 2e\ I_d G^2 N_d B$$

$$i_e^2 = 2e\ I_1 B$$

$$i_g^2 = \frac{2e^2}{hv} QP_g G^2 N_d B$$

$$i_b^2 = \frac{2B}{JW} \left(\frac{eGQP_0}{hv} \right)^2 \left(1-\frac{B}{2w} \right)$$

where N_d is the noise figure for the random avalanche process, N_t is the noise figure for the baseband amplifier, B is the information bandwidth, I_d is the detector dark current, W is the spectral width of the source, J is the number of spatial modes of the source and P_g

is the average background radiation power. The two dominant noise currents are i_q and i_T for systems with laser light sources. Since the avalanche gain process is seen to introduce gain-dependent noise, there is an optimal gain value for each given set of operating conditions.

VIII. Pulse Spreading

The degree of pulse spreading in the transmission lines effects the repeater spacing. If the pulse spacing or RMS pulse width is less than 1/4, the effects of dispersion are negligible on repeater design. For widths greater than 1/2 the intersymbol interference (ability to distinguish successive pulses) becomes severe and the receiver sensitivity rapidly degrades, with an increase in the system noise level. When the fiber impulse response is such that the pulse width is less than 1/4, the allowable transmission cable loss is the difference between the transmitted power and the receiver sensitivity and the system is considered loss limited. When the transmission cable pulse spreading is excess, with pulse widths >1/2, the cable length must be made shorter than the loss limited value to reduce intersymbol interference to an acceptable level and the system is considered to be dispersion limited.

IX. Power Consumption

Optical repeaters must be designed to have low power consumption to minimize operational costs while maintaining sensitivity to completely regenerate the received signal. To reduce power consumption the amplifier bandwidths in the repeater are restricted to be slightly greater than the desired maximum bit rate. The power consumption of the transmitter section of the repeater depends upon the threshold current of the light source (usually a laser for high data rate systems), the drive current and the source duty cycle. To minimize the power consumption of the overall fiber optic system beyond that of the repeater alone, an optimal light

source drive level must be used. This optical light source drive
level for long distance runs, represents the trade-off between the
individual repeater power consumption and the total number of repeaters used in the transmission system.

X. Repeater Spacing

The maximum spacing between repeaters is determined by many
factors including the dispersion characteristics of the fiber cable
and the electrical characteristics of the repeater unit. Given the
relationships between required receiver power for various data rates
the receiver dependent component of the repeater spacing is given
by:

$$S_{max} = (P_c - P_r)/A$$

where S_{max} is the maximum repeater spacing in Km, P_c is the power
coupled from the source to the fiber (dBmw), P_r is the required
optical power (dBmw) for a given BER, and A is the attenuation loss
of the fiber (dB/Km).

For low-loss cables presently available a reasonable attenuation figure is 5dB/Km. It should be noted from the equation to determine repeater spacing that the attenuation figure is in the de-
nominator. Since the values for the attenuation losses are low,
typically 4-8dB/Km, repeater spacing becomes a rather sensitive
function of the attenuation, A. For example, the difference of an
attenuation loss of 4dB/Km versus 6dB/Km would translate to a repeater spacing difference of 1.5Km which for long distance cables
can make a very significant impact on the total installation and
operational costs.

The repeater spacing is also dependent upon the fiber cable
dispersion characteristics. The greater the material and modal
dispersion of the fibers the shorter the repeater spacing will have
to be to maintain signal integrity (BER). The material dispersion
of optical fibers is given by

$$\sigma_{mat} = \lambda_o \frac{\delta\lambda}{c} \frac{d^2 n_1(\lambda)}{d\lambda^2} \ ,$$

where λ_o is the nominal source optical wavelength, $\delta\lambda$ is the RMS
spectral source width, and $n_1(\lambda)$ is the index of refraction of the
fiber core material. For a laser source, λ_o is typically 890nm and
$\delta\lambda$ is 20Å. For a LED source, λ_o is typically 810nm and $\delta\lambda$ about
370Å. For silica fibers n_1 is typically 1.5 and $\frac{d^2 n_1(\lambda)}{d\lambda^2}$ is 1.52 x
10^{31}ns/km^2. Thus, σ_{mat} with a LED is typically 1.6ns/km and 0.09
ns/km for laser sources. The modal dispersion for step index fibers
can be approximated by $n_1 \Delta/4c$ where $\Delta = 1 - (1 - (NA)^2)^{\frac{1}{2}}$. For graded index fibers the modal dispersion is approximately given by $n_1 \Delta/80c$.
For a fiber NA of 0.16, the modal dispersions of silica fibers with
$n_1 = 1.5$ would be $\sigma_{mod}(SI)$ = 16ns/km and $\sigma_{mod}(GI)$ = 0.8ns/km. The
total RMS pulse width contribution from both modal and material
dispersions is given by $\sigma = (\sigma_{mod}^2 + \sigma_{mat}^2)^{\frac{1}{2}}$. For a LED source, with
SI fibers, $\sigma \approx 16.1$ns/km and 1.8ns/km for GI fibers. For a laser
source with SI fibers, σ = 16.0ns/km and 0.8ns/km for GI fibers.
For step-index (SI) fibers the dispersion limitation on repeater
spacing is essentially independent on the light source due to the
dominance of the modal dispersion factor.

Using the typical fiber and light source values given above
Figures 25 and 26 show the repeater spacing as a function of bit
rate for dispersion limitations (SI and GI fibers) and for power
loss limitations (APD and PIN detectors with integrating bipolar
preamplifiers) for LED and laser light sources respectively. The
curves for the fiber dispersion limitations assume mode-mixing in
both the step and graded index fibers.

By 1980, low-loss fibers and cables will become more preva-
lent in optical communications systems. Where attenuation levels of
5dB/Km can be considered common in 1978, by 1980 3dB/Km will be common and by 1985 possibly 2dB/Km. The state-of-the-art cable at the
present time (December , 1977) is 2.7dB/Km with a total dispersion
of 10ns/Km. Figures 27 and 28 give the repeater spacings for LED

and laser light sources respectively using the best cable currently available. In the next few years in addition to attenuation changes the total coupled power into the fiber should increase slightly with improved sources and couplings which will also increase the maximum allowable repeater spacings.

In Figures 25 - 28 the BER for a single repeater is 10^{-9}. In systems with a large number of repeaters the total link BER can be much higher. For example a 5000Km transoceanic cable with a 5km repeater spacing would have a cumulative worst case BER of 10^{-6}.

FIGURE 25

MAXIMUM REPEATER SPACING VS. BIT RATE,

LED, A=5dB/km

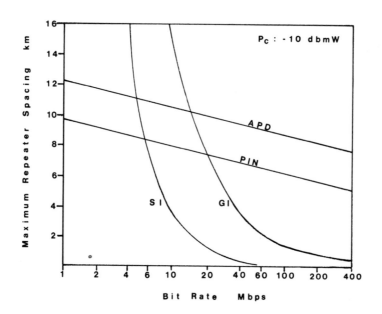

$\sigma_{mat} - 1.6 \, ns/km$; $(SI)\sigma_{mod} - 16 \, ns/km$;

$(GI)\sigma_{mod} - .8 \, ns/km$; $\lambda_0 - 870 \, nm$; $\delta\lambda - 370 \, \overset{o}{A}$

FIGURE 26

MAXIMUM REPEATER SPACING VS. BIT RATE,

LASER, A=5dB/km

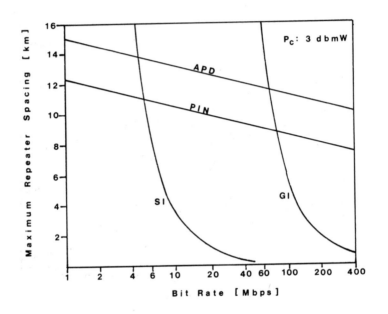

$\sigma_{mat} - .09 \, ns/km$; [SI] $\sigma_{mod} - 16 \, ns/km$;

[GI] $\sigma_{mod} - .8 \, ns/km$; $\lambda_0 - 890 \, nm$; $\delta\lambda - 20 \, \overset{\circ}{A}$

FIGURE 27

MAXIMUM REPEATER SPACING VS. BIT RATE,

LED, A=2.7dB/km

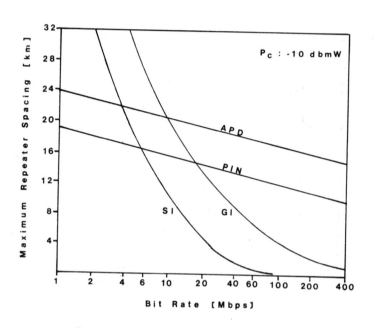

$\sigma_{mat} - .06\,ns/km$; (S1) $\sigma_{mod} - 10\,ns/km$;

(G1) $\sigma_{mod} - 0.5\,ns/km$; $\lambda_o - 840\,nm$; $\delta\lambda - 20\,\text{Å}$

FIGURE 28

MAXIMUM REPEATER SPACING VS. BIT RATE,

LASER, A=2.7dB/km

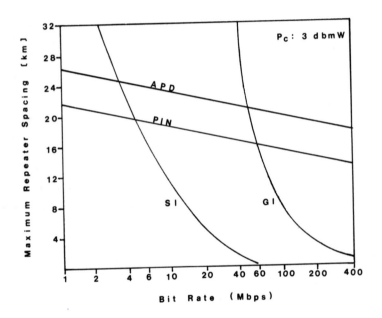

$$\sigma_{mat} - 1.2\,ns/km\; ; \quad (SI)\sigma_{mod} - 10\,ns/km\; ;$$

$$[GI]\,\sigma_{mod} - 0.5\,ns/km\; ; \quad \lambda_o - 840\,nm\; ; \quad \delta\lambda - 300\,\mathring{A}$$

References for Chapter 4

Photodetectors

1. Baack, C. et al., "GaAs MESFET: A High-Speed Optical Detector,"
   Electronics Letters, Vol. 13, No. 7, 1977.

2. Boisrobert, C.Y. et al., "Low Noise Photodetection: Preamplifier Analysis," Second European Conference on Optical Fiber
   Communication, September, 1976, Paris, France.

3. Caporossi, D. et al., "Design of Impedance Transformers for
   Increasing the PIN Photodiodes Applications Range," Ibid.

4. Ishii, K. et al., "Double Epitaxial Silicon Avalanche Photodiodes for Optical Fiber Communications," Electronics Letters,
   Vol. 13, No. 10, 1977.

5. Kondo, A. et al., "Frequency Response of Avalanche Photodiodes: Diffusion Current Dependence," Electronics and Communications in Japan, Vol. 58, No. 12, 1975.

6. Kondo, A. et al., "A Low-Noise Silicon Avalanche Photodiode,"
   Electronics and Communications in Japan, Vol. 58, No. 4, 1975.

7. Misugi, T. et al., "Detectors·for Optical Fiber Communications,"
   International Conference on Integrated Optics and Optical Fiber
   Communications, July, 1977, Tokyo, Japan.

Repeaters

8. Banks, F.M. et al., "An Experimental 45Mb/S Digital Transmission System Using Optical Fibers," Proceedings of the ICC,
   1974, Minneapolis, Minnesota.

9. Eppes, T.A. et al., "Low-Power Fiber Optic Repeater Design,"
   Electro-Optical Systems Design, May, 1977.

10. Goell, J.E., "Input Amplifiers for Optical PCM Receivers,"
    Bell System Technical Journal, Vol. 53, No. 9, 1974.

11. Gravel, R.L. et al., "Star Repeaters for Fiber Optic Links,"
    Applied Optics, Vol. 16, No. 2, 1977.

12. Mochida, Y. et al., "Low Noise 100Mb/S Optical Receiver," Second European Conference on Optical Fiber Communication, September, 1976, Paris, France.

13. Personick, S.D., "Comparison of Equalizing and Non-Equalizing Repeaters for Optical Fiber Systems," Bell System Technical Journal, Vol. 55, No. 7, 1976.

14. Uchida, T. et al., "An Experimental 123Mb/S Fiber Optic Communication System," Topical Meeting on Optical Fiber Transmission, January, 1975, Williamsburg, Virginia.

CHAPTER 5

SYSTEM DESIGN

I. Introduction

The design of fiber optic communications systems is determined and based upon four basic user requirements: the desired
data rate (bandwidth), the SNR (or BER), the distance between
terminals (or end-devices), and the type of source information
(digital or analog). Once these basic requirements are established
the design must consider numerous external variables, such as
physical and chemical environmental conditions, cost, reliability,
upgradability, size, weight, power requirements and effects of
EMI (electromagnetic interference) on transmitters, repeaters,
substations and end-devices. Many of the variables in optical
system design are interrelated. In some designs, some parameters
are pre-determined by the nature of the specific application such
as environmental conditions. All of the facts combined makes the
presentation of optical systems design rather complex if all
cases were to be considered. In this chapter, the basics of the
design process will be given along with specific examples typically encountered in fiber optic communication system design.

The first step in the design process usually is to determine if the dispersion bandwidth of the fiber cable and light
source combination are suitable to be used for the required data
rate (information bandwidth) at the given terminal or end-device
spacings. Since the dispersion bandwidth is a function of distance, and cable and light source parameters, it is possible

147

that in some cases the best available cables and lasers can not
meet the required data rates without using an inordinately large
number of repeaters which exceeds the alotted budget of the system.
If this is the case, ways can sometimes be found to shorten the
total link distance or to reduce the required data rates at minor
sacrifice to end-device service quality, thus permitting the use
of an optical system, within present budget limitations which can
be upgraded in the future.

The entire optical communications industry is in a large
state of growth and expansion. The quality and prices of cables
and light sources will undergo the greatest transitions in the
next five years, which brings up the question of upgradability.
Upgradability is a general term applied to fiber optic communications systems meaning the improvement of the available bandwidth
(data rate), extending the life-time and reliability of components
particularly the light sources, improving the attenuation of
dispersion properties of the cable or reducing the losses at
splices, connectors and couplings with new equipment or designs.

The optical communications systems being installed during
the period from 1977-1981 have difficult decisions to make in
terms of planning for upgradability of light sources, receivers,
transmitters, modulators and couplers, since the prices and
quality of future components can not be precisely determined.
Given that the cost of the initial purchase and installation of
fiber cable is rather significant in terms of total system cost
it is usually best to install the lowest attenuation cable
possible for applications where upgradability is a factor, such
that the original cable will not have to be replaced in the near
term. Cable is available now rated for less than 2.8 dB/km and
dispersions less than 11 ns/km. By installing LED light sources
now, in a few years, when low cost laser sources are available,
the system could be upgraded relatively economically and quickly.
Using slightly higher attenuation cables with lasers to attain
the required bandwidth often forces the future installment

of additional cable to meet expanded data role requirements.

In 1981 when low cost cables will be available, rated at
less than 2.5 dB/km, and low cost lasers will be available
rated for better than $5x10^5$ hours, initial design choices will
be much easier than they are during this transitional growth
stage of the industry.

Once a fiber cable and light source and modulation
scheme have been found that meet the bandwidth requirements,
giving consideration to upgradability, then the rest of the
system parameters must be determined if the desired performance
levels in terms of SNR and BER can be achieved. The major
parameters to be considered here are the coupling light source
power levels, fiber attenuation and dispersion at the source
wavelength and the receiver sensitivity. Most of initial market thrust will employ digital transmission where repeaters are
required. When short length analog links are used an important
parameter to consider is the level of sign distortion which
usually severaly affects performance quality in analog transmission.

Once the rough system design has been completed with a
set of specifications for all components, the next step in the
overall design process might be aptly described by iterative
perturbation and response analysis. Here, each separate component or system element is slightly altered (specifications
changed) to see what effect the small change will have on the
total system design and component choice. The procedure estab-
lishes a reasonable range of component specifications to meet
the job requirements within the given budget. It is not important where this iterative procedure begins since most of the
system elements are interrelated in terms of the final performance evaluation. If certain system elements have definite or
narrow ranges of choices available it is usually easier to

start with these elements in the design since much unnecessary analysis can be eliminated.

It should be evident that no "optimal" design is really achievable in a practical fiber optic system. The reason for this is that practical designs are severaly limited by the constraints of time and cost. Even if cost were not a constraint, an "optimal" system designed today using the best available components will be outdated (non-optimal) in a rather short period of time. The fiber optics communications industry will most likely undergo several transitions. The first transitional state is from 1977 to 1981 where new installations will be made and component availability and price will undergo rapid change. The next transition will be roughly from 1981 to 1986 where numerous installations will be made with low cost laser light sources and low attenuation fiber cables. The next transition or stage will be mostly a waiting period for integrated optical components to enter the field. The major concern for most companies is how well will the present design be adaptable to component changes in the time period when low cost lasers and low attenuation fibers are available. For some systems, such as underwater cables, the design must consider that upgradability of cables and repeaters may not be economically feasible and will at the very least be a very expensive proposition. Transoceanic cable is perhaps the best example of a system with limited economic upgradability potential.

II. Fiber Cable Choice

There are many commercially available fibers and cables for use in fiber optics. (See Table IV.) The number of companies making fibers and the lists of the transmission line products they offer is continually growing. Thus, the options in design would appear to be rather large. However, system limitation and requirements can rapidly narrow the choice of possible fibers

and cables. The first choice to make is whether to use single
fibers or fiber bundles. For single strand fiber cables a large
amount of power emitted by the source is lost in coupling the
fiber. Fiber bundles have larger NA values and have less losses
in coupling than single fibers. By using multifiber bundles
and lenses it is possible to reduce the coupling losses to be
approximately equal to the packing fraction loss. Over short
lengths where attenuation limitations are not critical, bundles
are often used if the required data rate is relatively low. For
long lengths, single fibers are usually used to make the cable
bandwidth capabilities as large as possible.

   Once the decision has been made to use bundles or single
fibers the next step is to calculate the total system losses
starting from coupling losses at the light source to coupling
losses at the end-device. For these calculations the following
cable parameters must be specified: effective fiber NA, index
profile, attenuation at the chosen light source wavelength, and
the length of cable between each repeater or terminal. When
fiber bundles are used the cross-sectional area and packing
fraction must also be specified. The total losses that can occur over the transmission pathway include losses in the fiber,
fiber splices, connectors and detector-source coupling losses.
To minimize input coupling losses the light source and detector
packaging designs should match the fiber or fiber bundles to
which they are attached. By the use of antireflection coatings
or index matching materials and properly matching the areas of
the components to the cable, the coupling losses can be kept
within an acceptable level for most communication system requirements. To determine whether or not the total loss is acceptable
depends mostly upon the receiver sensitivity as discussed in
Chapter 4.

   After an acceptable cable has been chosen on the basis
of system losses, dispersion limitations must be considered
to determine if the dispersion bandwidth is compatible with

the desired data rate. Once a cable or set of cables has been
found that meets the transmission requirements the environmental
and strength factors must then be considered. The major concerns
here are the required cable strength for the specific application,
whether or not internal conductors are needed for repeater power,
temperature range of the application, the effect of water on the
cable life-time and the total cable life-time under the anticipated environmental stresses. From these final considerations
a cable choice can hopefully be made. It will sometimes be the
case that no existing cable satisfied the system requirements.
If the system requirements can not be made less stringent, an
optical fiber cable transmission system is not feasible for
the given application until improved cables are available.
Using the lowest attenuation cables available today, most known
communication systems can be reasonably converted to fiber optic
transmission.

A flow chart of the fiber design choice process is given
in Figure 29. In the future, flow charts such as these for
each system element can be translated into computer programs
based on data banks of the existing commercial components. It
may then be possible to interlink the individual element flow
charts to obtain a totally computerized design based on existing
components and equipment. The major advantage of computer
designs is the ability to examine a large number of component
variations and modulation schemes.

III. Receiver Design

There is an ever increasing number of pre-packed receivers
for fiber optic links but most existing units have relatively
low bandwidth capabilities (<30 MHz) and are designed for indoor
applications. As a result, most receivers particularly for high
data rates are designed to accomodate a specific application.
Once the system requirements and constraints have been established
the first step in the design process is the decision to use

FIGURE 29

FIBER CABLE DESIGN FLOW CHART

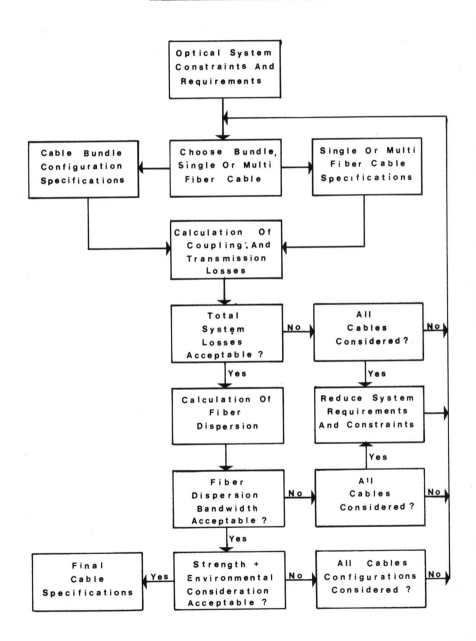

analog or digital modulation of the light source. Most of the
first applications and installations of fiber optic units will
use digital modulation. If digital modulation is used the next
step is to decide upon a particular modulation signal format such
as PCM or PPM. The next step for either modulation scheme is
to calculate the receiver noise equivalent bandwidth (NEB) which
is determined by the desired information bandwidth, the noise
introduced by various receiver components and the possible use
of filters to remove noise outside the transmissions signal
bandwidth. Once this is established the input SNR can be calculated from the BER and NEB.

The next step in the receiver design process is to select
a particular transmission wavelength from the existing light
sources available and a particular photodetector (PIN or APD).
Needlessly to say the source wavelength must be compatible with
the optical responsivity of the detector. Given that there are
numerous light sources, detectors and detector modules available,
this step in the design process has the largest range of possible
choices. In high data rate systems APD detectors are usually
chosen for laser light sources.

Once the light source and photodetector have been chosen,
the type of preamplifier must then be determined selected from
one of four types; resistive, loading, integrating (FET or bipolar) or transimpedance feedback. Hybrid combinations can also
be used. At low data rates the integrating FET preamplifier is
often used. At high data rates (<100 Mbps) integrating bipolar
preamplifiers are most often used. After the choice of the preamplifier has been made the next step in the receiver design
is to calculate the minimal detectable signal level which is given
by the minimal received optical power after coupling losses to
generate a current equal to the NEB noise current. The required
signal power to achieve the desired SNR is then calculated. If
this power level can not be achieved the photodetector selection
will have to be upgraded or else change the type of signal format

(for digital transmission) to make the required signal bandwidth
smaller.

Once the minimal received signal criterion has been
satisfied the next step is to consider the dynamic range of the
repeater.   The term "dynamic range" is meant to encompass a
variety of potential system variables including environmental
changes (particularly temperature), different path lengths
between repeaters or terminals, and changes in component characteristics with time. To be properly prepared a receiver design
must anticipate the worst component responses as a function of
potential component degradation particularly the light source
and detector and potential temperature excursions in the environment. If the anticipated temperature range has too large an
effect on the receiver, temperature compensation circuitry
(either precision control or feedback types) must be used both
on the receiver unit and on the light source unit as well.

Finally, the packaging of the receiver must be considered
which must account for environmental factors of temperature,
humidity and corrosion as well as protecting the detector from
possible stray light sources. In addition, some applications
will have to consider physical stresses applied directly to the
receiver such as low-frequency large-amplitude vibrations encountered in aircraft systems.

The flow chart depicting the receiver design process is
shown in Figure 30. The data bases for component choices and
the calculations involved in receiver design are much more
complex than those for fibers and cables. The cost of a repeater
unit at present is to a rather significant degree determined by
the choice of using a PIN or APD photodetector, thus upgradability considerations may weigh rather heavily for some borderline
designs.

FIGURE 30

RECEIVER DESIGN FLOW CHART

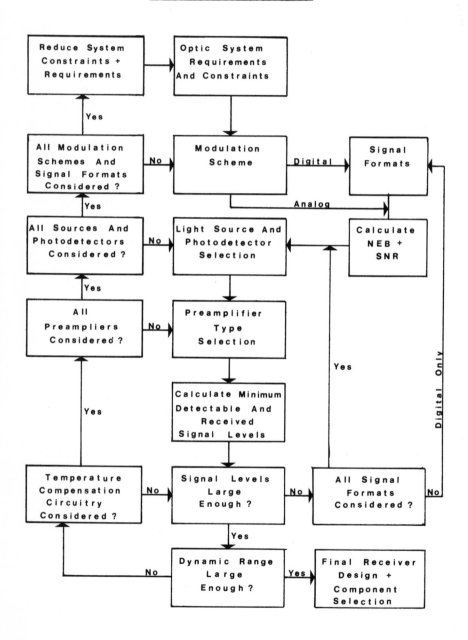

IV. Transmitter Design

There are about as many transmitters on the market as
receivers, designed for reasonably small bandwidth applications
(<30 MHz). Transmitters come in a variety of pre-packaged
forms some equipped with threaded socket connectors ready to
attach to a fiber cable, often in bundle format since most are
made for short links. The transmitter design centers basically
around the choice of an adequate light source and modulation
scheme.

The choice of the light source must consider the total
optical power required between terminals or repeaters, the output
wavelength and spectral width, the speed and linearity of response and the use of temperature compensation circuitry (if
needed). The choice of wavelength depends in part on the attenuation characteristics of the chosen fiber cable and in part on
the spectral response of the photodetector. The spectral width
depends mostly upon the restrictions on dispension bandwidth
within the transmission line. The required optical output power
is determined by the required received signal at the receiver
unit after considering and accounting for all of the transmission
and coupling losses.

The choice of source wavelength for silica and glass
fibers is usually between 840 and 900nm or in the 1060nm region
where fiber attenuations reach relative minima. For plastic
fibers used with short links a wider variety of wavelengths
can be used and the major consideration becomes required optical
power since attenuations in plastic fibers are so high.

The next step in the design process for optical transmitters is the choice of modulation schemes, digital or analog.
If analog modulation is used, the information bandwidth and
source power in terms of distortion levels must be considered.
If digital modulation is used the response speed of the light
source and choice of signalling format must be considered.
Once the modulation scheme has been determined, the next step is

to calculate all the coupling losses and determine whether or
not the optical coupled power is adequate. If it is not sufficient
a different signalling format could be tried for digital modulation or space division multiplexing could be considered.

Once the source and modulation have been selected several
parameters should be calculated and specified for design information on the receiver and fiber. These parameters should include
the transmitted SNR, required electrical power consumption and
the effects of temperature on the transmitted signal. If
temperature too greatly affects the output, then temperature compensation circuitry should be used on the light source drive current through suitable feedback circuitry or precise drive controllers.

A flow chart for the transmitter design process, similar to
that depicted for receiver design is shown in Figure 31. The chart
is again suitable for translation into computer format for use in
single or multi-element computer design.

Table XVII shows in a very brief tabular format the various
component choices for different ranges of link length or data
rate (bandwidth). The table is meant to show sample choices
of drivers, sources, fibers, detectors and amplifiers for three
basic groupings of fiber optical data rates. For intermediate
values, careful analysis must be done as outlined in the design
flow charts and in the following sections.

V Multi-Terminal Networks

Fiber optic communication systems can be used to distribute
various types of information signals to a number of remote terminals all inter-connected in a multi-directional data distribution system. These networks can be used for information transfer in aircraft, ships, computers, inter-building communications,
manufacturing and power plants and for cable TV systems. These
multi-terminal networks use the data bus system where a single
transmission liine can carry numerous multiplexed signals

FIGURE 31

TRANSMITTER DESIGN FLOW CHART

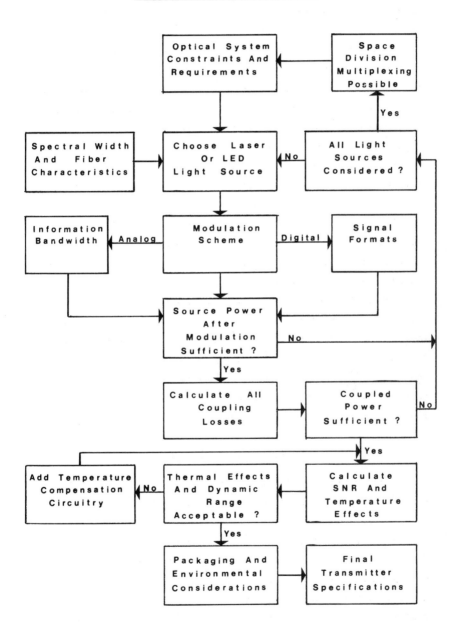

Table XVII

Component Choices for Different Data Rates

Component Type	Short Lengths < 40m Low Data Rates < 2 Mbps	Medium Lengths < 400m Medium Data Rates < 25 Mbps	Long Lengths > 1 km High Data Rates > 30 Mbps
driver	single I.C. or transistor	controlled current source	matched current source with pulse shaping
source	LED (Red,GaAs)	LED (red,GaAs) LED (GaAlAs,GaInAs)	LED (GaAlAs,GaInAs) DH Laser
fiber	plastic or glass high loss bundle	glass step index medium loss bundle or fiber	graded index low loss fiber
detector	phototransistor, PIN	silicon PIN	Si APD, GaAs APD
amplifier [a]	integrating FET	transimpedance	integrating bipolar

a - Typical preamplifier types are :

resistive   - voltage amplifier with low noise and resistive front end

integrating - integrating front end with bandwidth compensation circuitry

transimpedance - current feedback amplifier with low bias current

spatially distributed to a number of terminals and end devices.
The design of multi-terminal networks must consider the following.

1. The amount of optical power normally emitted from the
   original light source, the peak power available, the
   expected intensity degradation rate and the expected
   source life time.

2. The coupling bases from the light source to the transmission cable and from the cable to the various network components.

3. The required BER and SNR to maintain signal integrity
   and end-device quality.

4. The amount of optical power required to properly activate the photodetector (PIN or APD) and repeaters,
   if used, to maintain the desired BER and SNR.

5. The maximum possible system losses from coupling,
   splicing, repeaters, fibers, signal-splitters and
   temperature and degradation effects.

6. The most efficient and economical network distribution
   system for the particular application.

7. The total system upgradibility given future technological advances in light source intensity, speed and life-
   time, increases in repeater-amplifier bit rate capacity
   and decreases in system losses with new couplers,
   fibers, splices and signal splitters.

To estimate the amount of optical power available for
distribution of the data in the multiterminal system, a cable
and light source must be chosen. From the reflection and insertion losses, packing fraction, numerical aperture and fiber
diameters in the cable, and the radial directionality of the
light source, the amount of light coupled into the fiber can be
determined. For fiber bundles with large NA values this coupling
of optical power can be as high as +10dbm using laser sources.
For single fibers with relatively low NA values this coupled
optical power can be as low as -12dBm for LED sources.

VI   Data Bus Design

The two common types of optical distribution systems for communication systems are the series distribution system using access couplers and the parallel distribution system using star couplers. A system which combines both of these types of access coupling and signal divisions is called a hybrid configuration. Figure 32 shows a simple schematic diagram for an N terminal system for series, parallel and hybrid data bus design.

The alternatives in data bus design are quite numerous increasing rapidly with the number of individual transmitters, repeaters, sub-stations, terminals and end-device units, and must consider which units are to have the ability to transmit or receive information from the other units in the system. For example in a cable TV system the decision must be made as to whether or not the user terminal should be equipped to receive only now and later expand to an interactive system. The design, construction and performance of an optical data bus system depends on many factors including system transmission requirements, the number of present terminals to be installed, future expansion access, transmission line and coupling bases, environmental considerations and the maximum dynamic range needed between any pair of terminals.

The dynamic range (DR) in the star data bus for an N terminal system is given by: $DR = A_{avg}{}^{(2-N)/2}$ where A_{avg} is the average attenuation between terminals or stations given by $A_{avg} = C^{-aL/(N-1)}$ where a is the attenuation coefficient of the fiber of length L. For the series bus system the dynamic range is given by:

$$DR = \left\{ A_{avg} \; L_c{}^2 \; L_t \left(\frac{N-2}{N-1} \right) \right\}^{2-N}$$

where L_c is the connection insertion loss and L_t is the total transmission loss in the coupler given by $L_t = \frac{T_t}{1-2C_t}$ where T_t

FIGURE 32

DIAGRAMS FOR VARIOUS N-TERMINAL

DATA BUS DISTRIBUTION SYSTEMS

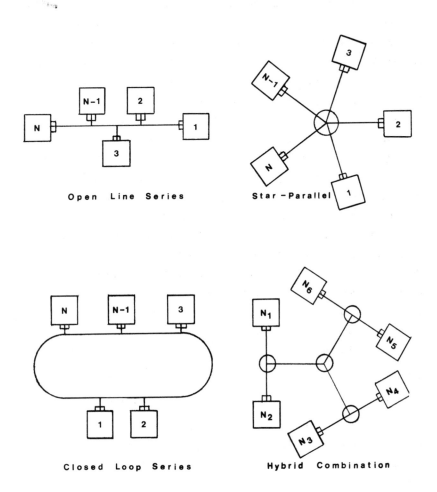

Open Line Series   Star-Parallel

Closed Loop Series   Hybrid Combination

is the fraction of power at the input coupled to the output of the coupler and C_t is the fraction of the total power removed from the bus. The average value for $L_t L_c^2$ is about 1.5dB.

The power ratio (PR) of the optical power in the transmission cable at the input to one terminal to the optical power at the output to another terminal is determined by the sum of the various system losses. For the radial system PR is given by:

$$PR = A_{avg}^{N-1} \, L_c^4 \, L_{is} L_s / N$$

where L_{is} is the star coupler insertion loss and L_s is the bidirectional coupler splitting factor usually ½ (3dB). For the series system the value of PR is given by:

$$PR = 0.5 \, A_{avg}^{N-1} \, L_c^{2N-2} \, L_t^{N-2} \, C_t \, (1-2C_t)^{N-2}$$

The total distribution system losses (TSL) in dB excluding fiber losses can be expressed as follows for the series network:

$$TSL = L_{sf} + L_{tr} + L_{ct} + 2L_{cl} + (N-3)(2L_{cl} + L_{ct} + L_{it})$$

where L_{sf} is the bidirectional splitting factor (3dB), L_{tr} is the tap ratio of the T-coupler (usually about 10dB), L_{ct} is the internal insertion loss of the T-coupler (usually about 1.5 to 2dB), L_{ci} is the connector insertion loss usually about (0.5 to 1.5dB), and L_{it} is the insertion loss from the T-coupler tapping given by 10 $\log_{10} (1-L_t)$. For the radial network TSL is given by: $TSL = 4L_{cl} + L_{cs} + L_{is} + L_{sf}$, where L_{is} is the insertion loss from the star couple tapping given by 10 $\log_{10} N$ and L_{cs} is the internal loss of the star coupler (usually about 6-8dB).

Using the typical values given above, the equation for the series TSL as a function of N is simply:

$TSL \approx 4.5 \, N + 3.5$. For the radial network: $TSL \approx 10 \log_{10} N + 13$. These simplified equations are useful to demonstrate the difference between using star and series networks. For $N \leq 5$ the total distribution system losses are about the same for both networks, but for a larger number of terminals the difference rapidly increases. For N = 10, the TSL for the series network

is about 48dB and for the star network TSL is about 23, or approximately a 2-fold loss difference. Figure 33 shows the total
system losses for star and series networks using the above parameters.

In some cases it may be required or desired to keep the
received detector signals approximately the same for all transmission sources. This can be accomplished by using a different
coupling factor at each terminal to provide a uniform response at
each detector. Here, the same set of coupling factors must be
used for one side of each coupler from terminal 1 to N as is
used   on the other side of each coupler from terminal N to 1.
This scheme can get to be rather burdensome as reflected in the
complexity of the power ratio given by:

$$PR = \frac{A_{avg}L_c^2 \left(A_{avg}L_tL_c^2\right)^{N-2}\left(1-A_{avg}L_tL_c\right)^2}{2 \quad \left(1-A_{avg}L_tL_c^2\right)^{N-1}}$$

This type of data bus system is sometimes called the
tapered duplex system. The dynamic range for the tapered duplex
data bus network is given by: $DR = 1 + A_{avg}L_tL_c^2$.   In the
radial system the power ratio varies as $1/N$ and as the exponent
of N for the series network. The tapered duplex data bus power
ratio is also not as favorable as the radial bus, as well as
having the disadvantage of being rather inflexible in terms of
future expansion since the tapered duplex system must be designed
exactly for N terminals.

The star (parallel) system has a signal level advantage
over the series system especially for a large number of terminals
(>10). Since the parallel system has only one mixer it does not
have the receiver problems associated with series systems which
must use large dynamic range automatic gain control devices to
process large signals from nearby terminals and small signals from
distant terminals. The disadvantage of the parallel system is
that the length of each terminal extension is rather large
meaning that more cable must be used to accommodate an equal

FIGURE 33

TOTAL DISTRIBUTION SYSTEM LOSSES

VS. NUMBER OF TERMINALS USED

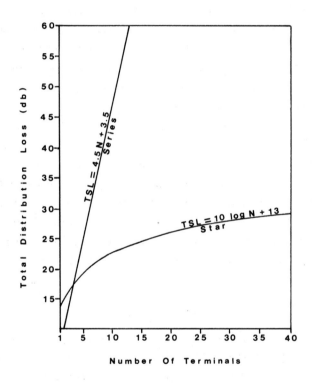

number of terminals than would be needed for a series system. It
should be noted that many applications do indeed use only a small
number of terminals in the region where the distribution system
losses for series and parallel systems are about equal. In these
cases the decision to choose either type of system must be based
on other factors such as cabling costs and weight and whether
or not future upgrading and terminal additions are planned.

The radial data bus configuration also has an advantage
over the series (T coupler) system in terms of dynamic range,
since power can be distributed uniformly to all terminals, making
dynamic range only a function of the distance to each individual
terminal. If the distances are all equal, assuming equal coupling
losses can be attained, the dynamic range in the star system
can approach the theoretical value of 1 (0dB) when cables of equal
attenuation are used. The disadvantage of using this system is
that the total length of cable is longer than used in the series
configuration. This disadvantage can in part be compensated for
by running a single multifiber cable part of the way between terminals and then splitting off the separate fibers to the individual terminals which is best accomplished by choosing a
centrally located transmitting substation to service a number of
nearby end-terminals.

Hybrid data bus configurations which can combine some
elements of all types of bussing configurations (radial, series
and tapered) are usually used for large networks. It is possible
in some hybrid data bus designs to approach the dynamic ranges
achievable in the radial systems with total optical power losses
that are only slightly larger. Using hybrid designs it is possible to achieve greater flexibility for future expansion and
alterations and use less total cabling than in the simple radial
system. In some cases such as large interactive cable television
networks, hybrid networks may be the only practical data bus
configuration.

VII   Simplified Design Procedures

The complete design of a complex multiterminal fiber optic
communications system can be very detailed and complicated since
the number of variables to consider are rather large. The decisions on cost limitations, payback periods (if any) and user requirements including bandwidth, SNR, BER, number of terminals
and terminal spacing are usually made on a managerial or corporate
level. The costs of large systems are often dependent upon cooperative mutli-company ventures which often makes simplified
rule-of-thumb cost evaluations rather inadequate or inapplicable.
The details of economics in fiber optic applications are discussed in Chapter 6.

Temporarily disregarding the financial considerations, the
basic engineering decisions to make on a user specified system
center about component choices: APD, PIN, LED, Laser, fiber
cable, repeaters, modulators, signal format encoders, multiplexers and demultiplexers. The two major factors to consider
are the optical power requirements and the response time requirements of the total system.

The optical power requirements and calculations are shown
in a rather simplified format in Figure 34. The first steps in
the calculation procedure are to specify the required bandwidth
or data rate (step 1A), the required SNR or BER (step 1B), the
signal format and number of terminals (step 1C), and the data
bus type and terminal spacing (step 1D). These are determined
mainly by the application specifications of the user but can sometimes be altered by the systems designer to optimize power distributions or to minimize cabling between terminals. The next step
is to choose an appropriate fiber cable (step 2), which can be
selected from Table IV or from the fiber product literature.
Actually, in the design process this step of the fiber choice can
be saved until the end of the calculation by merely assuming some

FIGURE 34

BASIC OPTICAL POWER CALCULATIONS

Information
Source

		Information Source
1A.	Required bandwidth = ___ MHz or Bit rate = ___ Mbps	Ap. Sp.
1B.	Required SNR = ___ dB or BER = ___	Ap. Sp.
1C.	Signal format : ___ , Terminal spacing L = ___ km	Ap. Sp.
1D.	Data bus type : ___ , Number of terminals N = ___	Ap. Sp.
2.	Fiber type : ___ , Attenuation A = ___ dB/km	Table IV
3A.	Light source type : ___ , Average output power ___ dBm	P.L.
3B.	If digital : NRZ add -3 dBm, RZ add -6 dBm	- - - -
3C.	If source output halfed to extend lifetime add -3dBm	- - - -
3D.	Total available light power : $P_s=3A+(3B)+(3C)=$ ___ dBm	- - - -
4A.	Detector type and manufacturer specifications: ___	P.L.
4B.	Required received optical power P_r = ___ dBm	Fig 35-37
4C.	Total power margin = P_s - P_r = ___ dB	- - - -
5.	Total fiber loss = $A \Sigma L_N$ = ___ dB	- - - -
6.	Source coupling loss = ___ dB	P.L.,E.D.
7.	Detector module coupling loss x N = ___ dB	P.L.,E.D.
8.	Number of splices ___ x Loss/splice ___ = ___ dB	Table VIII
9.	Allowance for temperature variations = ___ dB	Table XIX
10.	Allowance for component time degradation = ___ dB	Table XX
11.	# of extra connectors ___ x Loss/connector ___ = ___ dB	Table VIII
12.	Data bus attenuation allowance = ___ dB	Fig 33
13.	Total attenuation = 5+6+7+8+9+10+(11 or 12) = ___ dB	- - - -
14.	Excess power = EP = 4C - 13 = ___ dB	- - - -
	If EP ≤ 0 upgrade components or use repeaters	- - - -

```
Ap. Sp. = Application Specifications
P.L.    = Product Literature
E.D.    = Experimental Data
```

attenuation and rise time (dispersion) values and later determining if these assumed values were sufficient for the particular
design.

The choice of the light source is next in the calculation
procedure, which is usually specified by manufacture device number and the rated average output power in dbm (step 3A) which
can be found in the light source product literature or Table VII.
If the signal format in NRZ -3dbm is added to the output power
value and -6dbm for RZ formats (step 3B). In many cases it is
desirable to extend the lifetime of the light source to lower maintenance and replacement costs. To do this the peak light output
power is limited. In the calculation this is accounted for by
adding -3dbm (step 3C) to the specified average output power.
The total available light power P_s is then determined by the sum
of steps 3A, 3B, and 3C.

The basic choices of components of course depends on many
variables. To prevent excessive reiterations on the design procedure, Table XVII outlines same basic component choices depending
upon the bandwidth or transmission length of the given fiber optic
communications system.

The detector choice follows after the light source calculations. The type of detector used (APD or PIN) must be specified
by the manufacturer and device number which can be found in
Tables XI or XII or in the detector product literature (step 4A).
The required received optical power P_r can be determined from
the required data rate or bandwidth and SNR or BER from Figure 24,
35, 36 or 37 depending upon the signal format used (step 48). When
computer designs are used the questions to use in place of the
figures are given where applicable on the individual figures
and curves. Once the required received optical power has been
determined the total available power margin can be determined
(step 4C).

The next steps in the optical power calculations deal with
various fiber, coupling and splicing losses. First, (step 5)

FIGURE 35

REQUIRED OPTICAL POWER VS. BANDWIDTH FOR

ANALOG RECEIVER WITH PIN DETECTOR

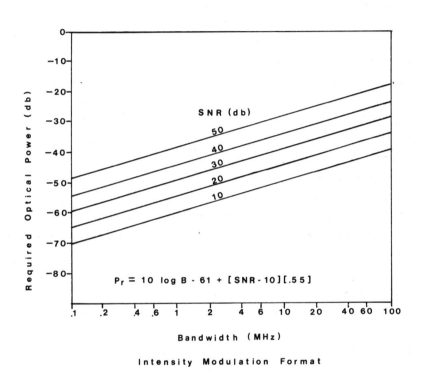

Bandwidth (MHz)

Intensity Modulation Format

FIGURE 36

REQUIRED OPTICAL POWER VS. BANDWIDTH FOR

ANALOG RECEIVER WITH APD DETECTOR

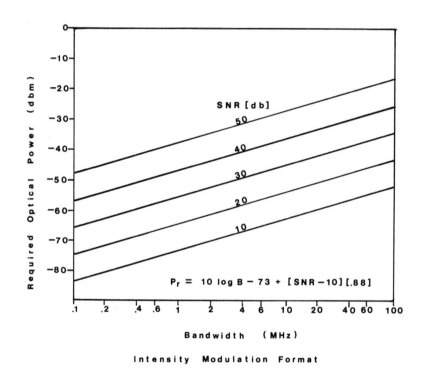

Bandwidth   (MHz)

Intensity   Modulation   Format

FIGURE 37

REQUIRED OPTICAL POWER VS. BANDWIDTH

FOR ANALOG RECEIVER - PPM&PCM MODULATION

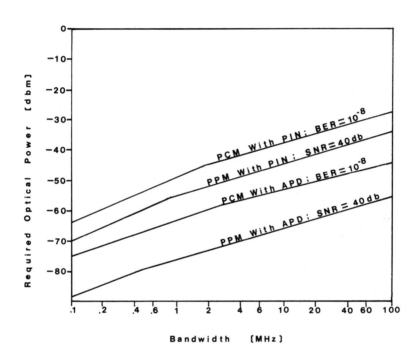

the total fiber loss is determined by $A \sum_{N-1}^{N} L_N$. If the system
has terminals which are all equally spaced the fiber loss is
given by $(A)(L)(N)$. The next loss to consider is from light
source to fiber coupling (step 6) which can be determined from
the product literature, experimental data or by calculation as
discussed in Chapter 2-VII. This loss is often an important consideration in the choice of fiber NA and whether or not to use
bundle or single fiber cables. The next loss to consider is the
detector coupling loss which can be determined from the product
literature, experimental data or by calculation (step 7). When
larger area devices are used this coupling loss is usually rather
small (<2db).

The next loss concerns cable splicing losses (step 8)
which may or may not be used in a given system. The use of
splices is often determined by the terminal spacing requirements
and the maximum available lengths of the chosen fiber cable. Table
XVIII briefly summarizes the expected splicing losses depending
upon the fiber type.

In any practical fiber optic link allowances must be made
for the degradation of components with time, particularly the
detectors and light sources. Allowances must also be made for
variations in the environmental temperature. In some indoor applications the variation of temperature is nominal and there is
little concern for allowances to compensate for detector sensitivity shifting and light source drive current changes. However,
for most outdoor applications temperature range is an important
consideration especially when no compensation circuitry is used
for the light source and detector modules. Table XIX lists the
allowances that should be made for temperature variations (step 9)
depending upon the temperature range and the use of compensation
circuitry. Table XX shows the loss allowances for component time
degradation (step 10) depending upon the light source-detector
combination.

Table XVIII

Expected Connector and Splice Losses

Fiber Type	Demountable Connector Loss(dB)	Fixed Splice Loss(dB)
Step	1.0	0.3
Graded	1.4	0.5
Bundle	2.0	1.0

Table XIX

Temperature Variation Loss Allowances

Compensation Circuitry	Temp.Range 10^0-30^0C	Allowance Loss (dB)
No	No	4
No	Yes	2
Yes	No	1
Yes	Yes	0

Table XX

Component Degradation Loss Allowances

Component Combination	Allowance Loss (dB)
LED + PIN	2
LED + APD	3
Laser + PIN	4
Laser + APD	5

If there are extra connectors in the fiber optic link
the losses from these various connectors are the next losses to
be accounted for in the optical power calculations (step 11).
Table XVIII summarizes the expected connector losses depending
upon the type of fiber cable used. For multiterminal networks the
total data bus attenuation must be considered (step 12). This can
be evaluated by independent calculation or by using Figure 33.
The equations for the series and star TSL can be re-evaluated
using the actual chosen component losses for signal splitters,
connectors, etc., to give a more accurate estimation of expected
network losses. Losses using the series network can become very
high above 7 terminals, necessitating the use of repeaters between
terminals.

The total system attenuation can then be determined (step
13) by adding all of the above individual losses. The excess
power EP, is then calculated (step 14) by subtracting the total
attenuation from the power margin. EP will often be a negative
value. If EP is between 0 and -20db the use of an APD instead
of a PIN detector, a laser instead of an LED light source, the
use of lower loss fiber cable or the use of a star versus series
networks can sometimes shift EP to a positive value meaning the
power requirements have been satisfied. If not, repeaters will
have to be used. For digital transmission the use of repeaters
is essentially limited only by total cost and BER considerations.
For analog transmission however the use of repeaters becomes
rather limited due to signal distortion. Often the designer will
come to the conclusion that for certain user applications there
will be no alternatives (in analog transmission) except to convert
the analog signal to some digital format and use repeaters to meet
the terminal spacing and SNR user requirements. Figures 25-28
show the expected repeater spacings using LED and Laser light
sources with APD and PIN detectors for various data rates with
low attenuation cables as discussed in Chapter 4. These figures

FIGURE 38

BASIC RISE TIME CALCULATIONS

Information
Source

1. Total allowable system rise time = ___   Table XXI
2. Light source (module) rise time = ___   P.L.
3. Photodetector (digital) rise time ___   P.L.
4. Receiver module (analog) rise time = ___   P.L.
5. Fiber : NA=___ , n=___   P.L.
6. Source : λ_o=___ nm , $\delta\lambda$=___ nm   P.L.
7. Modal dispersion rise time = TMOD = MD x L = ___   P.L. or
   SI : MD=600n$[1-(1-NA^2)^{1/2}]$, GI : $MD_{SI}/20$ Calcul.
   Series : TMOD=MD ΣL_N , Star : TMOD=MD x L_{max} or E.D.
8. Material dispersion rise time = TMAT = MT x L = ___   P.L. or
   MT= λ_ox $\delta\lambda$x 10^{-4} (SI or GI)   Calcul.
   Series : TMAT=MT ΣL_N , Star : TMAT=MT x L_{max} or E.D.
9. Rise time squares sum = S = $2^2+(3^2$ or $4^2)+7^2+8^2$=___   - - -
10. System rise time = SRT = 1.1$\overline{1}$ x $\overline{S}^{1/2}$ = ___ - - -
    If SRT > AST choose faster detector and/or   - - -
    light source and/or lower dispersion cable   - - -

P.L. = Product Literature
E.D. = Experimental Data
Calcul. = Calculation

---

apply for single links and do not account for the losses encountered in multiterminal networks.

Once the optical power requirements have been satisfied
the system rise time must be calculated to determine whether or
not the chosen components and fibers can respond fast enough to
handle the given data rate or bandwidth. The various steps in
the calculation of the rise time are shown in Figure 38. The
first step is to determine the total allowable system rise signalling formats (step 1). Table XXI shows the total rise time
in units of nanoseconds (ns) for various modulation and signal
formats as a function of the bit rate or bandwidth. For the PPM

Table XXI

Allowable System Rise Time Calculations

Modulation & Signal Formats	Total Rise Time (ns)
Digital NRZ	0.7/(bit rate)
Digital RZ	0.35/(bit rate)
Intensity Modulation	0.35/bandwidth
Pulse Position Modulation	$(SR \times PS/PW \times bandwidth)^{-1}$
Pulse Code Modulation	$(SR \times bits/sample \times bandwidth)^{-1}$

PS/PW = Pulse Separation to Pulse Width ratio (See Figure 39)
SR = Sampling Rate = 2.5 unless otherwise specified
    bandwidth units in MHz ; bit rate units in Mbps
    bit/sample = (SNR/6) + 1.2

---

format the total rise time is a function of the pulse separation
to pulse width ratio as shown in Figure 39. At higher SNR values
commonly encountered in optical systems the PIN and APD curves
in Figure 39 can be put into a simple equation form as shown on the
curves for computer design calculations.

Once the allowable system rise time has been determined
the response time of the light source unit must be specified
(step 2) from the light source product literature. Similarly
the response time of the photodetector or detector module is
specified (step 3). For analog transmission the response time
of the complete receiver module must be considered, usually determined from the product literature (step 4). The next step is to
calculate the modal and material dispersion rise times from the
fiber NA and refractive index values (n_{clad}) and the light source
emission wavelength (λ_o) and spectral width ($\delta\lambda$) as shown in
steps 5-8. The calculated fiber rise times (ns/Km) are multiplied

FIGURE 39

SNR VS. PULSE SEPARATION/WIDTH RATIO

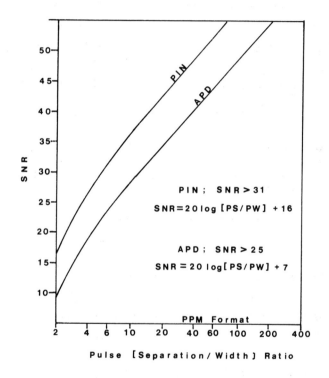

Pulse [Separation/Width] Ratio

by the appropriate fiber length between terminals depending upon the type of network used to link all of the user terminals. For star networks, L_{max} is defined as the maximum terminal-to-terminal distance in the system.

Once all of the individual rise times have been determined the sum of the squares of the rise times S, is then calculated (step 9). The actual expected system rise time (SRT) is then calculated (step 10). If the value of SRT is greater than AST which is often the case faster detectors or light sources will have to be used or else use cable with smaller dispersion values. It should be noted that steps 5-8 are only used as a means of approximating the dispersion rise times when they have not been determined experimentally. To be safe, especially with large networks, it is best to determine total cable rise time from experimental data or from the manufacturer's specifications after careful consultation with the product engineers to evaluate the accuracy and applicability of the given dispersion value.

To demonstrate the use of all the simplified design tables and figures the following example design is worked through on the optical power and rise time calcualtions for some typical user specifications.

VIII Cable TV Design Example

A color cable television station has a large network of users. A new installation is to be put in using fiber optic cables and transmission. The local network distribution is for 10 user terminals equally spaced at 0.5 km intervals (using a star data bus with a SNR of 40 with PCM and a BER of 10^{-8} with a bandwidth potential of 6 MHz). Note that the portion of the user requirements in parentheses are in part determined by the designer based on the user needs. A high SNR and lower BER (10^{-9} for example) could be used but for most color television signal applications it would be unnecessary.

FIGURE 40

COLOR TELEVISION OPTICAL POWER CALCULATIONS

1A. Required bandwidth = 6 MHz
1B. Required SNR = 40 dB , BER = 10^{-8}
1C. Signal format : PCM , Number of terminals N=10
1D. Data bus type : Star , Terminal spacing L=0.5 km
2.   Fiber type : GI , Attenuation A=3 dB/km
3A. Light source type : Laser , Average output power P=14 dBm
3C. Source output halfed to extend lifetime : Add -3 dBm
3D. Total available light power : P_s=(14-3) dBm = 11 dBm
4A. Detector type : APD module , 3 ns rise time
4B. Required received optical power P_r=-55 dBm
4C. Total power margin = (11+55) dBm = 66 dBm
5.   Total fiber loss = 3 dB/km x .5 km x 10 = 15 dB
6.   Source coupling loss = 10 dB
7.   Detector module coupling loss = 1 x 10 dB = 10 dB
8.   Number of splices 0 x .5 dB/splice = 0 dB
9.   Allowance for temperature variations = 1 dB
10. Allowance for component time degradation = 5 dB
11. Number of extra connectors 0 x 1.4 dB = 0 dB
12. Data bus attenuation allowance = 10 log 10 + 13 = 23 dB
13. Total attenuation = 15+10+10+0+1+5+0+23 = 64 dB
14. Excess power = (66-64) dB = 2 dB

The optical power calculations for this system are shown
in Figure 40. Steps 1A, 1B, 1C and 1D are user defined as shown.
The choice of fiber cable, light source and detector is usually
rather difficult to establish on the first iteration on the design.
For the example, the best cable and light sources have been chosen
to demonstrate the current component potentials. In this case it
will turn out that no repeaters will be needed on economic advantage which outweighs the extra expense of using laser light
sources and modulators and low loss cable and APD detectors.

The choice of fiber is a GI low attenuation cable with
strength member reinforcement suitable to use on poled cable
applications or underground conduits. The cable is available

conveniently in 0.5 Km lengths, thus eliminating the need of
splices. The light source for this application is chosen to be
a high intensity laser source operated at non-peak power output
(non-peak drive current) to extend source lifetime and lower
maintenance and repair costs (step 3C). The detector choice is
an APD prepackage receiver module with a response time of 3ns
(step 4A).

Using Figure 37 the required received optical power for a
PCM format at 10^{-8} BER is -55dbm (step 4B). This gives a total
power margin of 66db (step 4C). The total fiber attenuation
losses are 15db (step 5). The source coupling loss can be rather
large for a high intensity light source if alignment is not
precise. For this application a loss of 10db can be reasonably
attained. The detector coupling loss for each detector is about
1db, giving a total loss of 10db for the data loss system. The
use of temperature compensation circuitry is usually used for
color television applications especially if the analog transmission
is not converted to digital. The loss allowance in this case
using compensation circuitry is rather low (step 9). The use
of APD detectors and laser sources is the highest component loss
allowance combination in fiber optic systems (step 10). It is
possible that in the near future the time dependent nature on the
laser sources can be diminished to the point where instead of using
5db for the laser-APD combination a value of 1 or 2db could be
used. To be safe in any design both the best and worst case
combinations (existing and future) for each step should be carefully evaluated to better comprehend the multitude of alternatives
in the design.

For the star data bus the total distribution loss can be
approximated by 10log N + 13 for existing connectors and signal
splitters (step 12). The total system attenuation (step 13) is
64db giving an excess power margin of 2db, sufficient to permit
considering the design for detailed examination. The largest
contributors to the system losses are the data bus losses and

FIGURE 41

COLOR TELEVISION RISE TIME CALCULATIONS

	Rise Time (ns)

1. Total allowable system rise time =
   $[2.5x(40/6 + 1.2)x6x10^6]^{-1}=$ — **8.1**
2. Light source module rise time - Laser — 5
4. Receiver module rise time - APD — 3
5. Fiber :   NA=.25,   n=1.5 — –
6. Source : λ_0=840nm, $\delta\lambda$ =2nm — –
7. Modal dispersion rise time = (MD)(0.5+0.5) — –
   $MD=30n[1-(1-.25^2)^{1/2}]$ = — 1.4
8. Material dispersion rise time=$840x2x10^{-4}$x(.5+.5) — 0.2
9. Rise time squares sum = 25+9+1.96+.04 = 36
10. System rise time = 1.11x6 — **6.7**

the light source coupling. Little can be done to lower the star
data bus losses since each connector and coupling loss in the
system is given as about the best achievable losses to date. The
source coupling loss may be lowered by using intermediate transfer
optics (lenses and index matching fluids) and by carefully
matching the source area to the chosen fiber.

The rise time calculations are rather straightforward as
shown in Figure 41. From Table 21 the total allowable system
rise time is calculated to be 8.5 ns. The light source chosen has
a complete module rise time of 5ns (step 2) and the receiver module
has a rise time of 3 ns (step 3 or 4). With a laser source the
low loss, low dispersion GI cable has a very low material and
modal rise time. The length used in the calculations (L_{max})
is the maximum path length a signal would travel to get to an
end-device in this case a cabled color television, which is

(0.5km)(2) = 1 km. For data bus systems using unequal terminal
to coupler lengths the sum of the two longest lengths must be used
in the calculation (step 7 & 8) especially for interactive systems.

The sum of the rise times squared is 36 (step 9), giving
a calculated SRT of 6.7 ns (step 10) which is less than the allowable rise time. Since the light source, photodetector components and fiber cable combination are sufficiently fast in
response time at the given wavelength, the optic system is acceptable in terms of both rise time and optical power.

As stated previously this particular design has the problem
that the required optical power is borderline, meaning no future
expansion is possible without the use of repeater-amplifiers some-
where in the data bus lines since the best available components
(laser and APD) and low loss cable (3db/Km) were used in the design. Typically this local network would be part of a large
hybrid network where expansion is planned in terms of making
several or many additional unit installations in a small area, or
by overdesigning local area installations to have more capability
than currently needed.

IX Digital Link Design Example

To contrast the cabled television local network design,
the following example describes a typical digital link with a
bit rate of 20 Mbps where only one transmitting and one receiving
terminal are used. Such a link could be for computer, telephone
or satellite ground station applications. For this digital example the chosen light source is an LED with an APD detector.
The fiber cable chosen is one readily available with the characteristics of the cable designed for the T-1 Bell Telephone system.
The cable is strength member reinforced with up to 10 GI fibers.
The link distance for this case is 8 km. Given that the available
cable lengths are 2 Km, 3 splices are needed to couple 4, 2 km
lengths to make the total link distance of 8 km. It should be

FIGURE 42

DIGITAL LINK OPTICAL POWER CALCULATIONS

1A.   Required bit rate = 20 Mbps
1B.   Required BER = 10^{-9}
1C.   Signal format : NRZ , Number of terminals N=2
1D.   Terminal spacing L=8 km
2.   Fiber type : GI , Attenuation A=5 dB/km
3A.   Light source type : LED , Average output power P=3 dBm
3B.   Digital NRZ : Add -3 dBm
3C.   Source operated at peak power output : +0 dB
3D.   Total available light power : P_s=(3-3) dBm = 0 dBm
4A.   Detector type : APD , 3 ns rise time
4B.   Required received optical power : P_r=-59 dBm
4C.   Total power margin = (0+59) dBm = 59 dBm
5.   Total fiber loss = 5 dB/km x 8 km = 40 dB
6.   Source coupling loss = 10 dB
7.   Detector coupling = 1 dB
8.   Number of splices 3 x .5 dB/splice = 1.5 dB
9.   Allowance for temperature variations = 1 dB
10.   Allowance for component time degradation = 3 dB
11.   Number of extra connectors 1 x 1.4 dB = 1.4 dB
12.   Data bus attenuation allowance = 0 dB
13.   Total attenuation = 40+10+1+1.5+1+3+1.4 = 57.9 dB
14.   Excess power = (59-58) dB = 1 dB

noted that the total link distance in all applications is always
given as the actual cable length required to transverse the point-
to-point locations. The actual ground distance is usually more
than the link distance since cables must often run on poles or in
underground conduit where extra lengths are required to go around
bends, up and down poles and hilly terrains. When designing even
a simple link the surrounding territory between terminals should
be carefully studied to determine the best cable pathway. When
more than a simple link is involved, such as in all multiterminal
networks, the analysis of the local area should be made to determine the minimum total cabling design which includes the best place-
ment of the central transmitter, repeater or signal splitter. This
aspect of the design phase is often overlooked for its importance

especially in networks where future expansion is possible or planned or in networks which have a borderline optical power capacity.

The optical power calculations for this example are given
in Figure 42 and are rather straightforward. For this example
a demountable connector is made on the receiver terminal which
is often a desirable feature to permit easy relocation or servicing of the terminal end-device. It should be noted in all
designs that it may be possible to transmit a portion of the
total required data rate on one or several other fibers. For
example if several instruments are data bused together, the
individual signals can be multiplexed onto one fiber or each
signal can be transmitted on a separate fiber or any intermediate
combination of multiplexing and separate transmission can be
used. For high data rates (>30 Mbps) it may be desirable both
in terms of economics and safety to use several fibers for the
link. This increases the total fiber cabling costs but increases
the distances between required repeaters and allows the other
lines to continue to transmit information if one line or component
should fail. In this example if all 10 GI fibers where used in
the cable at the fiber 20 Mbps rate per fiber the total cable
data rate would be 200 Mbps. In many cases of course the division
of the total required data rate cannot be made and the optical
power and repeater spacing must be designed for a single fiber
link.

As shown in Figure 42, the 8 km distance can be satis-
factorily handled at the 20 Mbps data rate using the LED + APD
combination. If a laser source was used the link distance could
be increased to 10 km or the data rate capacity could be increased
for the given 8 km link distance.

The rise time calculations are given in Figure 43. For
this cable the dispersion characteristics have been measured
experimentally and do not have to be estimated by calculation.
The total allowable system rise time is 35ns which is rather

FIGURE 43

DIGITAL LINK RISE TIME CALCULATIONS

		Rise Time(ns)
1.	Total allowable system rise time (.7/20Mbps)	35
2.	Light source rise time - LED	6
3.	Photodetector rise time - APD	3
7.	Total fiber dispersion - experimentally measured	10
9.	Rise time squares sum = S = 36+9+100 = 145	-
10.	System rise time = SRT = 1.11x12.1	13.4

large but typical of digital NRZ signal formats. The actual system rise time is calculated to be about 13.4 ns which is well within the specified limit.

When the link distance is large enough to require re-peaters the design calculations become an iterative procedure where numerous component combinations (laser, LED, PIN, APD) and fiber cable characteristics (dispersion, NA, attenuation) can be chosen to meet the required design. In these cases the prices (in various quantities) of each purchased item must be known including any needed repeaters, multiplexers, splices and con-nectors, to determine the economics of each potential solution. The choice of which solution to use for the actual design is usually made on a managerial level based on upgradability po-tential, contract competition bidding and many other factors.

X. HI-OVIS Project

Cable television (CATV) has made rapid technological advances in the past few years and is now used worldwide in many different ways. At first the main purpose of CATV was to solve the problem of poor reception of television signals in mountainous areas and in large cities where skyscrapers often interfere with clear television reception. Then CATV was installed in areas where only a few channels were on the air to make it possible to see programs aired by television stations in nearby cities, thus expanding the available programming to the individual users. The next basic stage of development of CATV was to broadcast various types of information closely related to specific communities, which normally would not be a part of nation-wide programming. The next stage of development was the addition of specialized independent broadcast devices, keyboards, facsimiles and other information communication functions. This type of systematized programming provides retransmission of several television channels, as well as providing dedicated service for local community information and other information transmission services. This type of multi-purpose CATV is called coaxial cable information system (CCIS) which uses branch type coaxial cable for transmission.

The Japanese government and several of the leading communications companies have developed and prototyped the next stage of cabled television called the Visual Information System, which has more channels than CCIS and utilizes a computer controlled optical communication system. The prototype is the HI-OVIS (Higashi Ikoma Optical Visual Information System) which was established by VISDA, Visual Information System Development Association, in 1972 now under the leadership of Tashiwo Doko (Chairman), Katsumi Soyama (President), Confumi Yamaguchi and Saburo Iijima (Executive Vice Presidents), Takeo Otaka and Mashahiro Kawahata (Managing Directors). The aim of the association is to develop and spread the use of the complete visual information system. The prototype demonstration of

the system feasibility was completed in 1976. The program is
presently planning to expand service to transform the small scale
field trial into a large scale multi-community service network.
Dr. Masahiro Kawahata, currently the Managing Director of VISDA and
the HI-Ovis Project Center, has graciously provided the author and
editor with the following information about the HI-OVIS Project and
future expansions of fiber optic communications systems in Japan.

The purpose of the Visual Information System is to provide a
complete two-way communications service, including police, fire and
medical protection, multi-channel CATV and interactive and educational audio-visual programming. The main characteristics of the
system are multi-channel capability using fiber optic cable, two-way
communications for information exchange between station center and
the home, instantaneous individual channel and information select-
ivity by user control, and community adaptibility for local information.

The programming service for neighborhood localities conveys
news, events, shopping information, train schedules, etc., specifically related to small areas (towns), quite seperate from the typical nation-wide coverage. The programs are based upon information
selected and edited by the local inhabitants and system users. By
using home installed cameras and microphones, coupled with home keyboards and mobile broadcast centers, individual participation for a
wide variety of programs can be accomplished. In addition, home
polling services can be used to obtain preferred programming information and to express community interest in political issues and
community affairs.

The television request service available to each subscriber
enables the user to select any program by home keyboard operation.
A wide range of broadcast programs are available as well as movies
and educational programs. Video cassette recorders and automatic
cassette changers and program controllers are provided at the main
center. A still picture system is available to users to convey detailed information such as timetables, important phone numbers and

and addresses, hospital and medical information, movie and theater
guides, various events and shopping information and educational
programs. All combined, the system provides 30 user channels.

The original HI-OVIS system consists of three major sub-
systems. First is the head-end equipment including origination
hardware for broadcast and request services and the computer control
system to handle the various subscriber service requests. The
request services include special movies, still picture and alpha-
numeric data services. The second subsystem is the information
transmission system itself which is based upon a switched network
architecture in order to be able to handle the request aspects of
the various services being offered. A shared frequency division
multiplexed (FDM) architecture as commonly employed for conventional
systems does not have enough capacity to handle individual service
requests. In the original HI-OVIS installation a 16-fiber trunk
cable to a remote video switch was employed with a 12-fiber distribution cable serving five subscribers (one fiber each for upstream
and downstream communication per subscriber including a spare fiber
pair for future expansion or for replacement of broken or damaged
lines). The third subsystem is the subscriber equipment which nomi-
nally consists of a terminal device controller, a keyboard and a
standard television receiver. The subscriber subsystem can be expanded to include a camera and microphone for home or business
originated visual and audio information transmission. The original
costs of the home subsystem for a limited number of subscribers was
relatively high, which was to be expected for the innovative equipment. However, with mass production techniques the cost as with all
fiber optic system components, will drop in the next few years. The
basic home unit for this project should be available for several
hundred dollars by late 1978 which coupled with the increased services available makes the overall system cost effectiveness and
value-in-use superior to coaxial systems. Home installed computer
systems can also be added to provide banking services, home
controlled television games and various traditional computer

services.

   In the expansion program for the HI-OVIS Project, there will
be three main optical trunk cables each containing 36 fibers connecting the head-end to a subcenter. From each subcenter, containing a video switch and associated control equipment, there will be
up to 14 distribution cables radiating out to various subscribers
each of which can serve 12 subscribers with 24 fibers per distribution cable. The final subscriber drop will be comprised of a two
fiber cable eminating from an optical junction box where it is
connected to the distribution cable. Figure 44 shows the detailed
schematic diagram of the various component systems in the HI-OVIS
Project. Independent optical fiber cables will also be connected to
local points such as schools and hospitals so that service programs
may be originated at these points and then transmitted to the head-
end for distribution to subscribers, similar to the distribution
method used for the mobile studio broadcasting system.

   VISDA is already in the process of expanding HI-OVIS, the
world's first and largest multi-service interactive fiber optic
cable television communications network. The largest interactive
CATV networks in the United States are coaxial systems, but by the
early 1980's optical cabled television will appear in many nations
designed and modeled to various degrees after the successful
Japanese field demonstration project which encompasses most every
conceptual and equipment aspect of projected fiber optics communications network systems.

FIGURE 44
HI-OVIS PROJECT SCHEMATIC

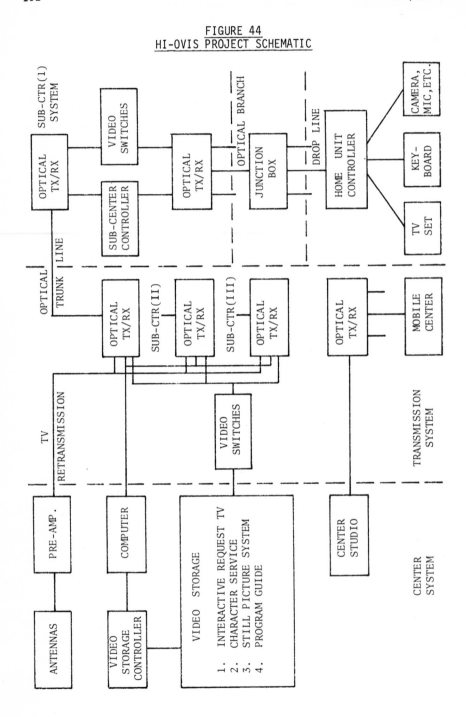

References for Chapter 5

Systems and Installations

1. Akiyama, S. et al., "3-Channel TV Transmission on Optical
   Fiber", International Conference on Integrated Optics and
   Optical Fiber Communications, July, 1977, Tokyo, Japan.

2. Aoki, F. et al., "Practical Use of Optical Fiber Communications for Electric Power Companies", Ibid.

3. Boerner, M. et al., "Single-Made Transmission Systems for
   Civil Telecommunications", Proceedings of the IEEE,
   Vol. 123, 1976.

4. Bouillie, R. et al., "Influence of Baseband Frequency
   Responses of Optical Components on the Transmission System
   Design", Topical Meeting on Optical Fiber Transmission,
   February, 1977, Williamsburg, Virginia.

5. Brackett, C. A. et al., "Optical Data Links for Short-Haul
   High Level Performance at 16 and 32 Mbps ", Ibid.

6. Callan, R. R. et al., "A Duct Installation of Two Fiber
   Optical Cables", IEE Conference Publication, Vol. 132,
   1975.

7. Daikoku, K. et al., "A Proposal on Optical Fiber Transmission
   Systems in a Low-Loss 1.0-1.4μm Wavelength Region", Optical
   and Quantum Electronics, Vol. 9, No. 1, 1977.

8. Eppeo, T. A. et al., "Long Distance Optical Communications,"
   Proceedings of the International Conference on Communications,
   1975.

9. Hiramatsu, T. et al., "An Experimental 400 Mbps Digital
   Transmission System Using Selfoc Fiber", Topical Meeting
   on Optical Fiber Transmission, February, 1977, Williamsburg,
   Virginia.

10. Kimura, T. et al., "Considerations on Optical Fiber Trans-
    mission Systems and Problems Associated with Their System
    Design," Review of Electrical Communications Laboratories,
    Vol. 24, 1976.

11. Markstein, H. W., "Fiber Optics for Electronic Interconnec-
     tion", Electronic Packaging and Production, April, 1977.
12. Miki, T. et al., "Hybrid Digital Transmission System over
     Optical Fiber Cable", International Conference on Integrated
     Optics and Optical Fiber Communications, July, 1977,
     Tokyo, Japan.
13. Nawata, K. et al., "Studies on PCM-IM 100 Mbps Optical Trans-
     mission Experiments", Electronics Communications Engineers
     of Japan, Vol. 75, 1975.

CHAPTER 6

ECONOMICS AND APPLICATIONS

I Introduction

Fiber optic systems must provide competitive performance
at a lower total price or provide improved performance (bandwidth,
SNR, reduced interference problems, etc.) at an equivalently competitive total price. In the evaluations and comparisons of fiber
optic systems to existing wire systems many business factors must
be considered including end-device unit cost, total installation
costs, operational costs, upgradibility, reliability, field service availabiility and maintenance costs. In addition, performance and environmental factors must also be considered including
system data rate (bandwidth), total power requirements, quality of
transmission, overall safety, size and weight, and resistance to
various applied stresses, vibrations, temperature fluctuations,
humidity, corrosive or explosive atmospheres and electro-magnetic
interference. In special applications such as computers, sur-
veillance and security and military systems, additional considerations may sometimes be appropriate including probability of failure,
down-time, undesired accessibility potential and resistance from
intruder attacks.

II Business Opportunities

The various types of business opportunities available in
the electrooptics market are rather numerous and basically depend
upon the size (capital potential) of the company. The five general

195

types of business ventures available now and in the near future
may be summarized briefly as follows.

1. Developing new components (detectors, repeaters, modulators, cables, lasers, LED's, connectors, splices signal splitters, multiplexers, etc.), or complete systems (multiterminal networks, interactive cable television, etc.) for production and direct sales to customer market.

2. Developing new components or complete systems to sell to other corporations or governments who will in turn sell to the customer.

3. Production of support equipment and materials for the production of components such as furnaces for fiber manufacture, cabling equipment, fiber dopant chemicals, plastic coatings, strength member materials, etc.

4. Providing installation and maintenance and repair service to users or suppliers, or a service to test, rate or to certify various fiber optic system components.

5. Production of auxiliary and field testing apparatus such as fiber cutters or welders, portable optical power meters and laboratory equipment for fiber and cable mechanical and optical testing.

III   International Economics

The major fiber optics applications market areas are in
telephone and non-telephone common carriers, cable television,
computers, industrial automation, ground based satellite stations,
and military applications. The major centers for installations
and production of fiber optic communications (signaling) equipment are the United States, Japan, Canada, France, West Germany
and the United Kingdom. By the eraly 1980's the United States
is expected to be producing more than one-half of the fiber optic

systems, and the Far East and Western Europe sharing the rest of
the market about equally.

The total signaling market (a subset of the communications
market) where fiber optics has the great potential to capture
significant portions of the market can be projected for the next
ten year period as shown in Figure 45. The part of this market
expected to be captured by fiber optic systems and components is
also shown on the graph using the 1977-1987 time period. By
1987 the fiber optics capture should amount to more than 14% of
the total signaling market equivalent to more than 1.2 billion
dollars by conservative estimates. This figure could approach
the 1.8 billion dollars mark if the telephone industry moves
faster or on a larger scale.

The breakdown by general categories of the curves shown
in Figure 45 are given in Table XXII. The telephone industry is
of course the largest worldwide market followed by industrial
automations and military applications. All of the figures (tabu-
lated numbers in Table XXII) are only for the United States, Canadian, Japanese, United Kingdom, West German and French markets
combined. The last category given in Table XXII is for "Others
Combined" which is an encompassing area including such areas as
lightning resistant instrumentation, satellite ground station
links, analytical instrumentation, medical equipment, etc.

During this period of growth and rapid expansion in the
fiber optics industry, the prices of components and cables are
expected to drop markedly with increased production volumes and
international competition. Figure 46 shows the projected costs
of fiber cables and connectors and laser sources over a ten year
period. The cable curve is for low loss dual fiber reinforced
cable. The connector curve is an average price for precision
single and multiple fiber connectors. The curve for light sources
is an average cost for DH lasers with a spectral width of about
20Å and lifetimes greater than 10^5 hours. It is evident from
this graph that the major drop in fiber optic communications

Table XXII

Projected Signaling and Fiber Optics Markets

General Area	Signal Equipment Market		
	1978	1982	1986
Telephone Industry	1.5	3.0	4.5
Industrial Automations	0.5	0.7	1.1
Military Applications	0.4	0.6	0.8
Cable TV Networks	0.09	0.11	0.13
Computer Applications	0.03	0.05	0.08
Others Combined	0.01	0.05	0.11
Totals (Billions $)	2.53	4.46	6.72

	Captured Fiber Optic Market	
	1982	1986
Telephone Industry	160	410
Industrial Automations	80	210
Military Applications	40	120
Cable TV Networks	15	25
Computer Applications	10	30
Others Combined	10	30
Totals (Millions $)	315	825

FIGURE 45

FIBER-OPTICS & SIGNALING MARKET

PROJECTED GROWTH RATES

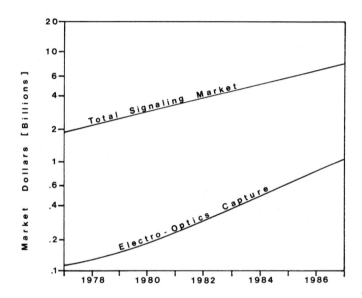

components occurs by 1982. Beyond this time prices will mainly
be determined by price competition among large companies, while
during the first few years prices are determined mainly by increases in production levels of all components and cables.

The cost of producing coated optical fibers depends upon
which method is used to prepare the fiber preforms and to draw
the preforms into fibers. Table XXIII shows the production costs
of making coated optical fibers by various methods on the basis
of dollars per coated fiber kilometer. The calculations for the

FIGURE 46

PROJECTED FIBER OPTICS

COMPONENT PRICE CHANGES

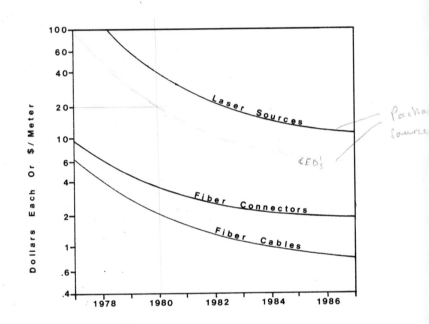

figures in this table were made on the basis of a large fiber
facility producing about 120,000 km/year of 120μm diameter
fibers on the basis of 1978 U.S. dollars. The major difference
between the various combinations of preform production and fiber
drawing methods is that production facilities using laser controlled fiber drawing will have to spend significant amounts of
fixed capital to purchase and safely maintain the laser unit.
Note that significant price changes occur depending upon the
materials yield on the overall process.

Table XXIII

Coated Fiber Production Costs by Different Methods

Cost Source ($/km)	CVD+Res.	CVD+Laser	SM + Res.
Fiber Raw Materials	0.25	0.25	0.15
Coating Plastic Materials	0.10	0.10	0.10
Labor And Overhead	6.85	6.85	6.85
Plant Utilities	0.35	0.40	0.45
Maintenance Supplies	1.50	2.10	0.90
Other Financial	1.45	2.20	1.35
Total Cost (100% Yield)	10.50	11.90	9.80
Total Cost (70% Yield)	15.00	17.00	14.00
Fixed Capital (Millions $)	0.90	2.00	1.00

CVD = Chemical Vapor Deposition Method
Res.= Using resistance heating for drawing
      fibers (versus RF induction furnace)
SM = Stratified Melt fiber process

IV Fiber Optic Applications

There are numerous applications of fiber optic communications systems and fiber optic links. Many of these potential applications have already been implemented to varying degrees and
others have yet to be utilized. The major application areas are
in telecommunications (telephone and telegraph), industrial automation and military applications. In the area of data services
fiber optics can capture a large portion of the future expanding
market which includes applications in banking and insurance systems
(multiterminal computer links), electronic mailing systems (commercial and government), transportation services, manufacturing

systems, news communications, and retailing and publishing applications.   In the area of visual displays fiber optics is highly
suited for computer plotters, news service photographic transmissions, micrographics, picturephone and videophones, educational
television links, paid television systems (such as in hotels and
in motels) and various medical display systems. One market of
great potential for fiber optic systems is for large multiservice
interactive cabled television networks requiring large bandwidths
as dicusses later.

Detailed descriptions of the major application areas are
given in the following sections.

V Telephone Applications

Using fiberoptic telecommunications systems solves some of
the problems inherent in wire systems, particularly ringing,
echoes and crosstalk for audio applications (phones). Transmission
even in high noise environments with fiber cables avoids interference and induced errors into the transmission. Lightning
damage to cables and connected equipment which are in outdoor
environments can be eliminated in some cases by using fiber optic cables with single pair internal power lines, instead of using
conventional multipair metallic conductors.

The applications of fiber optics into the telecommunications
area include voice telephones, video phones, telegraph services,
various news (wire) services, message services and data networks
(such as end-devices coupled to stock and commodities computers)
all transmitted over common carrier links.

Most telephone networks can be classified into three different levels of transmission based on the cabled distance between
terminals or exchanges and risers. These three levels are usually
given by local-loop plant (or simply local loop), short-haul
trunks (or simply short-haul) and long distance links. In the
next ten year period most of the fiber optics applications in the
telephone area will be for short-hauls except in undeveloped areas

where long-distance lines either do not exist or do not have sufficient capacity to handle existing or near future transmission
traffic./ Short-hauls of less than 12 km, which includes the vast
majority of switching exchanges, are most favorable to fiber optic
links since the distance can be made without the use of repeaters.
Immediate application potential exists in areas of Canada and South
East Asian and African nations where telecommunications links are
very often non-existent, antiquated or insufficient to meet current
demands. Opportunities are also available in many places for submarine telecommunications cables to link islands, countries and
continents.

  Fiber optic systems are cost competitive with the existing
copper cable line rising LED or laser sources, and especially if
multiplexing techniques are used. The cost per two-way voice
channel per kilometer of transmission line is about $1.30 for wire
cable systems with 8000 voice channels per system and $1.00 for
fiber cable systems with 16,000 voice channels per unit system.
The cost of copper greatly influences conventional wire cables.
The cost of fiber cables is greatly influenced by production levels
and demand which is expected to increase at a dramatic rate in the
next few years.

  The fiber optic cable designed by General Cable for the
T-1 phone carrier system is typical of what can be expected to be
used for some of the first telephone short-haul links. The cable
contains up to 10 fibers with an outer diameter of 2.8 cm and a
weight of 745kg/Km. The fibers with a 0.21 numerical aperture
have a 62.5µm core diameter and a 125µm cladding diameter. The
power conductor is a 22 AWG uncoated annealed copper wire pair.
Additional internal strength members are provided to give a maximum recommended pulling tension of 400 kg, a minimum recommended
bending radius of 30 cm and a maximum compression strength of
2400 kg/m. The reeled unit lengths of this mass produced cable
are 1 km or longer. The multimode graded index fibers of the
first cables used for the T-1 links have an attenuation level of

about 5db/Km. Future cables can be expected to have even lower
alternation values (3 to 5db/Km) with optical dispersions of
about 10ns/Km. Even this first generation cable rated a 5db/Km
provides superior repeaterless service for links of 10 Km or less
for the T-1 PCM bit rate of 1.544 Mbps.

The largest market for common carrier lines and equipment
is in the United States, which has the largest communications market in the world. The degree of capture (or penetration) of the
fiber optics systems depends to a large degree upon the equipment
developed and manufactured to interface fiber optic links to
existing wire systems and by the momentum put into manufacturing
and purchase of fiber optics components by ATT, the largest single
users of telecommunications equipment. In Europe the major market
thrust will come from France where the largest growth rate in
telephone signaling equipment is expected (15 to 20% over the
1978-1982 period).

The signaling market in the near future is heavily based
on digital transmission (signal format). However, in some local-
loop and short-haul applications analog transmission may still be
used where signal quality and distortion are not a limiting factor.

Japan is expected to have the second largest world market
in fiber optics equipment. One factor which has given a major
push for fiber optics links in Japan is that there is little
available space for additional ducting in most of the major cities.
Fiber optic cable can replace existing lines of fit into overcrowded areas providing more capacity and occupying less space in
overcrowded ducts or on existing common carrier above ground lines.
Japanese companies often have cooperative business ventures for
new technology with the government. The combined financial re-
sources and efforts should make Japan the world's leader in fiber
optic technology and may rival the United States in total production of fiber optic equipment and world wide systems installations.

VI   Interactive Cable TV

One of the most attractive aspects of cabled television
is the possibility of using interactive systems which provide
additional services to the users of multi-channel cabled color
television. ⅋ New systems have already provided continuing interactive sercice for smoke detectors and heat detectors to automatically alert fire departments of potential or active fire hazards,
police alarms to alert mobile vehicles of potential or active
crimes or traffic accidents, and medical alert alarms to summon
first aid or rescue squads for immediate medical aid and hospital
transportation.   The users in towns employing interactive cabled
television systems (such as The Woodlands in Texas) enjoy a complete protection and service system that do not appear to be
matched by any individual consumer service.

Towns or local areas currently employing interactive cabled
television systems have proven records of greatly reduced criminal
incidents (particularly burglary), very few damaging fires and
the fastest medical and police response times.   Automation of
services is accomplished by coupling all user interactive devices
(detectors, alarms, etc.) to a central computer which registers
the alarm signal and automatically alerts the appropriate agency
(fire, police or medical) to the nature of the problem and the
name, address and phone number of the individual household.

Interactive cabled television networks can also provide
additional services including automatic meter readings for light,
gas and electric companies and oil level readings to aid delivery
companies in servicing customer oil reserves.   These two-way
cabled networks can also provide user polling services and a means
for the station to automatically register the use of different
channels and services at any given time.   In more sophisticated
interactive systems the station can provide the individual user
with special educational courses (on tape) upon request.   In some
cases these course tapes can have interactive computer evaluations
of the users' progress in an individual subject by evaluating

prepacked test programs. This is essentially adding computerized
educational courses to the cabled television network. In addition,
computerized games such as tennis and hockey which are popular
as individual add-on television units could be provided upon user
request. Using a central computer system could provide the user
with more advanced computerized games such as chess, bridge and
other games. In such advanced interactive networks the bandwidth
or data rate requirements can become rather large necessitating
a larger number of individual fibers to be cabled to the individual
user terminals. These larger interactive multi-service networks
will not be on the market to any significant degree until about
1982 after laser sources, APD modules and low loss. ($< 3db/Km$)
multifiber cables are available at their projected low costs.

Multiservice cabled television systems offer much more
than simply a cabled version of atmospheric television transmission,
and should see an ever-increasing portion of the television market in the years to come. Fiber optic cabled television has a
vast advantage over coaxial cabled transmission lines especially
for large interactive systems due to their large bandwidth multi-
channel capabilities.

An example interactive system is the 30 channel system
being used by Warner Cable. This interactive network allows
viewers to participate in television programs using a push button
home selection unit. A central computer scans, registers and
tallies viewer responses for rapid screening (in the order of
six seconds of delay time). The system currently being installed
and tested includes game participation shows, various television
games and educational-type tests. For this particular system
customer charges are partly based on actual time usage of movies
and other programming instead of the equal-charge method for all
users regardless of usage. Other interactive systems have special
charges assessed for each time the customer uses (activates)
various alarm systems or local emergency services and special

charging rates based on total individual consumer usage of all
channels and services.

The market and development of cable television systems
varies dramatically throughout the world. In Europe there are
very few cable television networks and few under development.
The existing systems mostly use SDM access to a limited number of
channels. In Japan there are few networks in operation but a large
development program, the HI-OVIS Project (Higashi Ikoma Optical
Video Information System) is underway to bring multi-service
interactive cable television systems onto the market. Canada has
the largest percentage of household usage of cabled television
in the world (about 2/5ths of the homes) mostly with conventional
multi-channel (10 to 20) usage with few interactive networks currently available on the market.

The United States has a large number of cabled television
networks linked to over 12 million households the majority of
which are conventional (non-interactive wire) systems. A few and
increasing number of interactive networks are available in the
United States predominantly finding their way into new housing
developments. The United States and Canadian networks usually
use FDM to access different channels. Since the United States
and Canadian networks already have rather large cable television
equipment investments in wire cable links, the major thrust in
the fiber optics capture in this market will be in installing new
cable links into existing systems and for total fiber optic
systems for new networks. In some areas when large multi-service
interactive systems are developed a significant market incentive
may be available to replace or transform some existing networks
into interactive links.

In areas outside the United States the growth, development
and implementation of cable television systems depends to a
significant degree on governmental regulations, financial support
and/or political influences. Here, the international Fiber Optics

Communications Program of Arthur D. Little, Inc. may play a major
role in the development of fiber communication systems and worldwide optical telecommunications.

VII Industrial Automation

The fiber optics applications in industrial automation are
very diversified. The market can be divided into three basic
categories: process controls, discrete manufacturing automation,
and transportation and energy applications. The process controls
market basically consists of the chemical, nuclear, and petrochemical industries. Additional process control markets exist in
other industries as well, such as foods and metals manufacturing
facilities. The signal market in process controls was about 170
million dollars in 1977 and is expected to grow to 240 million
dollars in 1982.

The discrete manufacturing automation market includes
numerically controlled machines in manufacturing plants and job
shops and large factory data systems. The signaling market
here was about 120 million dollars in 1977 and is expected to
grow to 200 million dollars in 1982. The transportation and energy
category includes such applications as airway, shipping, highways,
railways, mining and electric, gas and oil transmission and distribution control. The signaling market in this category was about
$100 million dollars in 1977 and should expand to $160 million
dollars in 1982. The information rates are typically rather low
for all of these applications inthe range of 10^2 to 10^5 bps.
The major virtues of fiber optic systems over wired links for
industrial automation center around the freedom from electro-
magnetic inteferences, improved signal quality, large upgradability
potential and the lack of conducting wires for hazardous (explosive
or flammable) atmospheres (including dusty atmospheres in some
food manufacturing facilities). Fiber transmission lines used to
control remote equipment are free from problems of electromagnetic
noise particularly those caused by motors and relays. In addition

in some critical process applications fiber optic links can
easily be made with reserve or auxiliary power supplies to operate
when major system power failures occur.

Fiber optic systems can be expected to capture a significant
portion of the industrial automation signal market especially for
new manufacturing facilities. A change in the manufacturing standards for hazardous products could give fiber optic process control system a rapid push onto the market.

VIII Computer Applications

The computer signaling equipment market is rather small
compared to the other potential market applications. Unlike most
of the other application areas there are some communications applications in computers, mostly between internal components that
require very high data rates up to several Gbps. Auxiliary equipment such as plotters, card readers,. etc., require relatively low
data rates which can be handled both by wire and fiber links.
However in computer systems a major difficulty in interfacing
computer components to the CPU or the terminal modem is the interference by electromagnetic noise which causes errors and wasted
time in retransmission. In most coaxial systems this problem is
limited by reducing the data rates and cable lengths. Fiber optic
links are immune to induced noise irregardless of the data rate
or cable length even when properly designed repeaters are used.
For many computer applications fiber optics offers these advantages of freedom from electro-magnetic interference and
grounding problems and improved signal quality (lower BER's).
Most computer links are less than 3km and would thus not require
the use of any repeaters. Some existing and future computer
systems involve much longer links to the remote terminals to the
main computer such as remote links to time-sharing computers.
For these links repeaters have to be used and fiber optic systems
offer and advantage over coax or twist pair shielded cable links
in using fewer repeaters. /

The computer field in general is a very fast-paced equipment industry with rather large turn-over rates in replacing old equipment and upgrading existing systems. The major capture of the computer signaling market by fiber optic systems will most likely not occur until beyond 1986 when inter- and intra-computer links can be made using complete fiber optic component systems. The largest growth rate in the computer signaling marketing area is expected in Japan but the largest single market area will remain in the U.S. with about 65% of the worldwide market.

IX Military Applications

There are many possibilities for fiber optics communications in the military area including systems for aircraft, ocean surface vessels (ships), submarines, missile applications and various ground communications links. For the military aircraft applications commonly used by the Navy and Air Force possiblities exist for/data busing in every type of aircraft including attack jet aircraft, aloft command control centers, space vehicles and automated and strategic air command craft. For the ocean surface vessel applications commonly used by the Navy, Marines and Coast Guard, possibilities exist for inter- and intra-ship communications, submarine cable links, submarine mobile command centers, underwater missile stations and ship-to-satellite communications links. For the ground communications applications possibilities include missile center links, field commandpost links, ground-to-air communications, ground-to-ocean vessel communications and automated mobile surface missile command posts.

The major advantages of fiber optic systems considered by the military are the absence of crosstalk, radio frequency interferences, grounding problems and outside jamming influences, resistance to on-board electromagnetic interferences, the small size and weight of cables and equipment (especially for aircraft), the enhanced security level provided by optical transmission versus coaxial or wire pair cables, and the large upgradability potential (large bandwidths and data rates at low BER and high SNR).

For aircraft applications the advantages of fiber optic data busing include having a lighter plane and thus more payload capacity or range. The more an aircraft weighs the more it costs to operate and maintain. The savings over the life-time expectancy of aircraft per pound of excess weight eliminated is about $1000/lb. By using fibers instead of twisted shielded pairs the weight savings ratio is about 10 to 1. For typical military aircraft the weight savings using fiber optics ranges from 70 lbs. for the A-7D or A-7E to 370 lbs. for the B-1 (or similarly large future bombers).

It is appropriate here to comment upon the security nature of fiber optic transmission in comparison to conventional wire cables. For mobile transmission equipment such as in aircraft and submarines fiber optic systems can be considered superior to coaxial cables in terms of overall security and freedom from outside influences. For permanent ground installations fiber optic links offer better security and protection from outside unwanted "taps" than coaxial or twisted pair shielded cables. However, from the work developing for integrated optic networks it is clear that it is technically feasible to "tap" signals from optical fiber transmission lines by stripping a small portion of the plastic coating on the cladding and amplifying optical signals from the fiber cladding surface to obtain legible SNR levels. For high security systems however, by using various modulation schemes on irregular schedules with multiplexing and scrambled signal transmissions even the best fiber "tapping" schemes would be useless. For simple standard telecommunications links "tapping" should be possible without breaking the fiber line. If the fiber integrity can be temporarily broken, careful splicing coupled with the use of carefully optically energy balanced repeater-amplifier units makes "tapping" of fiber transmission lines a reality using existing technology and equipment for digital formats. For analog

transmission this sort of line interruption is presently very
difficult to attain due to the difficulty of returning the signal
to the line without additional distortion or slight lessening
of the signal quality. For analog transmission signal formats
the only way at present to remove information from the line without
detectability is with high level amplification of signals removed
from the cladding surface. Note that in all optical systems if
too much of the fiber coating or cladding is removed, the total
signal level may be noticeably diminished to a detectable level
especially in links with borderline optical power.

The United States military market level is expected to be
about $350 million dollars in 1978 and $480 million dollars in
1982. Fiber optic capture in this market area is more uncertain
than in the non-military applications since equipment and overall
budgeting decisions must be made by both military and congressional
personnel.

Existing examples of unclassified military applications of
fiber optic systems include the United States Navy's A-7 aircraft
which has undergone extensive aloft testing (fiber optic data
bus), the 60 analog channel sonar system for ocean surface vessels
(from the Naval Underwater Systems Center), the 31 digital channel link for command-control systems for destroyer escorts (from
the Canadian Navy), the 20 Mbps ground satellite station-data
processing link at Fort Meade in Maryland (from the United States
Naval Systems Center) and the 1.2 km rocket engine testing site
link at the Air Force's Arnold Engineering Development Center in
Tennessee. These examples are but a few of the existing military
applications of fiber optics communications links.

X Smaller Marketing Applications

There are numerous smaller volume marketing applications
for fiber optic communication systems. Perhaps the largest of
these is satellite ground station and radar applications. Both
military and commercial satellite ground station links may become

a significant market by 1981. Several fiber optic installations
in this area are already in operation and more are in the process
of being built or designed. By 1985 this signaling market could
grow to be about $75 million dollars. Fiber optic links could
easily capture a large percentage of this market which usually
requires data rates in excess of 15 Mbps.

Another area here is for analytical instrumentation and
various medical equipment employing fiber optic links. Fiber
optic bundles for exploratory medical examinations are already
quite common and there is ample room for expansion and improvement
in this small market area. Most sophisticated analytical instrumentation involves some data busing especially x-ray, Raman,
infrared, chromatography and spectroscopy apparatus many of which
interface with minicomputers.

XI Underwater Cables

Underwater cables are an excellent application area for
fiber optic systems. For underwater telecommunications links
fiber optic cable can transmit more information with less unit
weight and volume than conventional wire cables such as transoceanic cable. For short island links (<12 km) no underwater
repeaters are necessary. Thus no internal electrical conducting
elements are needed. Especially for these communications links
fiber optic cables offer a large advantage over wire systems.

There are some cases in military applications where is is
desirable to run underwater cables that have minimum detectability.
The location of most ocean transmission lines takes advantage
of the presence of copper inside the cable or the presence of
minute electric fields surrounding the entire cable line or the
large metal volume present at repeaters lying on the ocean floor.
Fiber cables with high tensile strength can be made with woven
plastic filament strength members with no internal metal components. This makes the cable undetectable but requires

self-contained power supply units for the repeaters. Using RTG
(radioisotope thermoelectric generators) sufficient power (0.5W)
over a long period of time (10-20 years) can be generated. By
encasing the repeater-power supply unit in thick plastic gels
and burying each unit beneath the ocean floor surface, fiber cable
links can be made that are virtually undetectable to outside in-
truders.

Another smaller market for underwater fiber cables is for
tethers of various sorts. Tethers are usually ship links to
submerged equipment such as underwater cameras or ocean surface
links to underwater laboratories. Fiber optic tethers are
desireable over wire tethers since the information bandwidth
for equivalent physical parameters can be made to be much larger
to accommodate high quality real time signal transmission.

XII Process Control Systems

Numerous industries and manufacturing facilities throughout
the world rely upon various types of process control instrumentation which may be manually or automatically operated by man or by
computer. Depending upon the nature of the process and the manner
and speed with which a process responds to change and correction,
the type of control system best suited may require simple on-off
(bang-bang) controllers to proportional plus integral (reset) plus
derivative (rate) controllers. There are numerous variables which
may be controlled depending upon the process or machine operation,
such as temperature, pressure, liquid and solid levels, flows of
gases, liquids and solids, densities of gases, liquids and solids,
viscosities of fluids, moisture and the chemical composition of
materials. Typical process operations in which control is important
include heating, cooling, distilling, absorbing, electroplating,
reacting, power generation, and machine tool manipulations. Controllers which regulate the various process variables come in many
different types including direct acting, floating, integral (reset),

proportional, derivative (rate), ratio, reverse acting, sampling,
self-operating (regulators), and time proportioning and scheduling,
and various combinations of all of these.

Various types of control mediums are used to interface servomechanisms, controllers and control systems (computers). These
include pneumatic (where air pressure acts as the control system
signal), electric (where voltage or current acts as the control
system signal) or hydraulic (liquid actuation). The servomechanisms
used in the process industries may be classified as positional or
velocity, analog or digital, proportional or on-off, translational
or rotational, direct-drive or gears, direct or alternating or
pulsed current, absolute value or incremental motion, or electric,
hydraulic, pneumatic or mixed.

With such a wide variety of controllers and servomechanisms
signals must very often be converted from one medium or one format
to another, requiring numerous forms of process control transducers.
For process control systems which will employ fiber optic links,
transducers must be available to convert electrical or mechanical
signals at the process site to light signals (pulses) into the
fiber cable. In addition, transducers must be available to convert
optic transmissions into electrical or mechanical signals for
meters, digital readouts, control indicators or for digital or
analog electrical formats for computer controls.

The general trends in the instrumentation field are to
attain greater accuracies and sensitivities, capability to measure
extreme values with applicability under extreme conditions of use
and capability of resolving or responding to changes or physical
effects which occur at extremely high speeds. The trend for wider
use of instrumentation systems for automated processes is continually growing, limited mainly by cost considerations, safety and
reliability. The entry of fiber optic links into the process
industry centers mainly around safety. The major safety benefits
of fiber optic transmissions in manufacturing facilities is the
elimination of wire cables to link servomechanisms, controllers and

computers. In wire systems, stray noises can often be introduced
into the line by nearby equipment which are eliminated by using
glass or plastic fiber cables. The servomechanisms and controllers
which interface most directly (easily) to fiber links are electrically based instrumentation. To achieve the fastest process response
a minimum amount of intermediate transducers (signal transforma-
tions) should be used. For facilities such as nuclear power plants,
the fastest control systems can be achieved using electrically and
magnetically based servomechanisms and controllers, linked with
fiber optic transmissions.

In processes where control systems can not shut down without
fear of catastrophy, it is necessary to include battery powered
back-ups to drive the fiber optic components. For control systems,
the light sources used will be LED's and the detectors will be
PIN's to maintain low installation costs for the low data rates
used. This makes fiber optic control instrumentation cost competitive with existing wire systems. In some cases where only short
distances have to be transversed, plastic fiber bundles can be
used, but for longer distances such as is required in larger petrochemical plants, low-loss fibers will have to be used since link
distances can sometimes exceed several kilometers. For almost all
types of fiber optic process control instrumetnation repeaters will
not be needed, thus direct analog transmission (without changing to
digital format) can be used.

The market for fiber cabled process equipment is expected to
grow slowly during the 1980's. The major oil, nuclear and power
generation companies may change to fiber optic based systems sooner
than other industries since they stand the most to gain by improved
safety and reliability. Changes in the federal safety codes for
manufacturing industries may spur activity in the area. Activity
may also increase if an insurance savings can be realized for
facilities employing the new technology.

The major business opportunities in process control and
machine control instrumentation are not in fiber cables or optical

components. Rather the major opportunities are in the manufacture,
sales and installation of inerfaces, transducers, controllers and
servomechanisms specifically made to be coupled by fiber optic
links. With so many companies in the instrumentation field, the
competition can be expected to be very stiff. Market dominance
could only be achieved by developing and presenting completed and
installed systems.

References for Chapter 6

1. Coryell, C. et al. "The Application of Optical Waveguides
    to Army Communications", Proceedings of SPIE, Vol. 63,
    1975.

2. Dworkin, L. et al. "Progress Toward Practical Military Fiber
    Optic Communications Systems", Topical Meeting on Optical
    Fiber Transmission, February, 1977, Williamsburg, Virginia.

3. Electro-Optic Communications - Phase II, Arthur D. Little,
    Inc. November, 1976.

4. Elion, H. A., "Fiber Optic Communication Needs of Developing
    Countries", Intelcom '77, October, 1977.

5. Elion, H. A., "Cost Effective Fiber Communication Systems
    and Markets", Proceedings of Third Electro-Optics Conference,
    1977.

6. "Fiber Optics For All Environments", Laser Focus, August,
    1977.

7. Fisher, J. et al., "Applications of Integrated Optics to
    Chemical Spectroscopy", Topical Meeting on Integrated Optics,
    January, 1976, Salt Lake City, Utah.

8. Gallawa, R. L., "Component Development in Fiber Waveguide
    Technology", Proceedings of Intelcom '77, October, 1977,
    Atlanta, Georgia.

9. Gallawa, R. L. et al., "Telecommunication Alternatives
    with Emphasis on Optical Waveguide Systems", Office of
    Telecommunications Report 72-73, Washington, D.C., 1975.

10. Hawkes, T. A. et al., "Optical Communication Systems for
     Aircraft", Second European Conference on Optical Fiber
     Communications, September, 1976, Paris, France.

11. Holmes, L., "A Total Fiber Communications Experiment In-
     volving Voice, Video and Data", Intelcom '77, October, 1977.

12. Kawahata, M., "The Interactive Cable TV Project of Higashi
     Ikoma", Proceedings of Third European Electro-Optics Con-
     ference, 1977.

13. Montgomery, J. D., "Worldwide Business Opportunities in
     Fiber Optic Communications", Fiber and Integrated Optics,
     Vol. 1, No. 1, 1977.

14. Popou, Y. M., "Applications of Optical Communication Lines
     in Computer Techniques", International Conference on Inte-
     grated Optics and Optical Fiber Communication, July, 1977,
     Tokyo, Japan.

15. Wolf, H. F., "Overview of Optical Fiber Systems Markets",
     Proceedings of Intelcom '77, October, 1977, Atlanta, Georgia.

APPENDIX

Glossary of Symbols

A_{avg}	average attenuation in optical fiber line
AM	amplitude modulation
ADM	adaptive delta modulation
APE	adaptive predicting encoding
b	bit rate [bit interval = (bit rate)$^{-1}$]
B	information bandwidth
BER	bit error rate
c	velocity of light
C	photodiode capacitance
CATV	cable television
CCTV	closed-circuit television
CVD	chemical vapor deposition
CVSD	continuous variable slope delta modulation
dBm	decibels relative to 1 mW (same as dBmW)
DCD	digital controlled delta modulation
DFB	distributed feedback
DH	double-heterostructure (junction laser)
DM	delta modulation
DPCM	differential pulse code modulation
e	electron charge
F	thermal noise
FM	frequency modulation
FSK	frequency shift keying
G	avalanche gain of photodetector
h	Planck's constant
i_b	source bias current
i_b^2	mean square beat noise
i_d	detector photocurrent
i_d^2	mean square dark current
IF	intermediate frequency
i_l	photodetector leakage current
IM	intensity modulation
i_n^2	mean square noise current after avalanche gain
i_q	quantum noise current
i_t^2	mean square thermal noise current
J	number of spatial modes seen by photodetector
K	Boltzmann's constant
L	fiber length (usually expressed in km)
LED	light emitting diode
LID	laser injection diode
n_{core}	fiber core refractive index
NA	numerical aperture
N_e	photodiode excess noise factor
NEB	noise equivalent bandwidth
NRZ	non-return to zero (for digital formats)
P_c	average transmitted optical carrier power

Glossary of Symbols (cont.)

P_o	average received optical power
P_r	average received optical carrier power
PAM	pulse amplitude modulation
PCM	pulse code modulation
PDM	pulse duration modulation
PL	polarization modulation
PM	phase modulation
PPM	pulse position modulation
q	electron charge
R	equivalent load resistance of photodetector
RZ	return-to-zero (for digital formats)
SNR	signal-to-noise ratio (dB)
T	noise temperature or sample of bit period in PPM
TDM	time division multiplexing
TNEP	total noise equivalent power
VSD	variable slope delta modulation
$\delta\lambda$	spectral width of source
α	attenuation coefficient of optical fiber line
ε	photodetector responsivity (R)
λ	wavelength (usually expressed in nm or um)
σ_{as}	applied stress to optical fiber
σ_G	standard deviation of avalanche gain process
σ_{max}	maximum local stress applied to optical fiber
σ_t	standard deviation of thermal noise
η	photodetector quantum efficiency (Q)

Table XXIV

Communication Systems Bandwidth Requirements

System Type	Receiver Equipment	Transmission Data Or Media	Avg. Required Channel Band.
facsimile	printer or display units	alphanumeric or graphical data by wire,coax or micro.	3 KHz-4 MHz
telegraph	typewriter or tape punches	alphanumeric data by wire or micro.	0.15 KHz
telephone	telephone or special terminals	voice or data by wire,coax or micro.	3 KHz-2 MHz
television	cathode-ray tubes	coax or microwave	6-15 MHz

Physical Constants

Quantity	Symbol	Value
speed of light in vacuum	c	2.997925×10^8 msec^{-1}
elementary charge	e	1.602189×10^{-19} C
Planck Constant	h	6.626176×10^{-34} JHz^{-1}
	$\hbar/2\pi$	1.054589×10^{-34} Jsec
Boltzmann Constant	k	1.380662×10^{-23} JK^{-1}

Conversion Factors

To Convert From	To	Multiply By
Angstrom units	centimeters	1×10^{-8}
	microns	1×10^{-4}
	millimicrons	1×10^{-1}
degrees	minutes	60
	radians	0.017453
meters	feet	3.280840
	inches	39.370079
	miles (nautical)	53.995680×10^{-5}
	miles (statute)	62.137119×10^{-5}
sq. millimeters	sq. cm.	1×10^{-2}
	sq. inches	1.55×10^{-3}
	sq. meters	1×10^{-6}
steradians	solid angles	79.577472×10^{-3}

Basic Definitions

aperture -- the diameter of the largest entering beam of light that can enter the optical fiber

attenuation -- the optical power loss per unit length along a waveguide which is the sum of the absorption and light scattering

bandwidth -- the complete range of frequencies over which a particular information system can function, basically determined by the maximum decodeable frquency

bend loss -- a type of increased attenuation caused by permitting high-order modes to radiate from the side of the optical fiber when the fiber is curved around a restrictive radius of curvature or when small distortions are introduced inside the fiber from manufacturing imperfections

cladding -- material usually of a low refractive index that surrounds the central core of an optical fiber to minimize surface scattering losses

coherent radiation -- light propagation where the phase between any two points in the field is exactly the same or else maintains a constant difference throught the duration of the lightpulse

core -- the center of the optical fiber bound by the cladding which conducts the transmitted light in the high refractive index region of the fiber

critical angle -- the smallest angle of incidence at which total reflection will occur at the boundary between two media of different indices of refraction

crush strength -- the physical limit of an optical fiber or cable to withstand an applied force or weight perpendicular to the axis of the fibers

dispersion -- mode and material dispersion in optical fibers causes a broadening of the input pulses along the length of the fiber, limiting the bandwidth by the degree of pulse-spreading

dopants -- chemical elements added to the core and cladding of fibers to alter the transmission and reflective properties of the fiber

electro-optic detector -- any device capable of detecting transmitted light by converting the received radiation into some form of electrical signal

Basic Definitions (cont.)

germanium detector -- a type of photoconductive detector
where germanium doped with other elements acts as a semiconductor in
the range of 600 to 1100 nm

graded index -- a type of optical fiber where the refractive
index of the core decreases radially outward towards the fiber cladding achieving high-bandwidth capacity and coupling efficiencies

injection laser diode -- a semiconductor device where lasing
occurs within at least one PN junction, where light is emitted from
the edge of the diode

junction diode -- the basic element of an injection laser
where the semi-conductor diode has the property of essentially conducting current in one direction

Lambertian emitter -- an optical light source where the
transmitted radiation is distributed uniformly in all directions

light emitting diode -- a PN junction semiconductor device
that emits incoherent light when biased in the forward direction
from the junction-strip edge or from its surface

modal dispersion -- the component of pulse spreading caused
by differential optical path lengths in a multimode optical
waveguide

multiplexing -- the combination of several information signals from different channels into a single optical channel for
increased bandwidth transmission

numerical aperture -- the measure of the degree of light
acceptance of a fiber defined by $NA = (N_{core}^2 - N_{cladding}^2)^{1/2}$
where N is the refractive index

optical cement -- a permanent and transparent adhesive, usually epoxy or methacrylate capable of handling extreme temperatures

optical waveguide -- any material structure capable of
guiding radiation along a path parallel to its axis containing the
light within its boundaries or adjacent to its surface

packing fraction -- the ratio of the active fiber core area
`otal cross-sectional area of the fiber bundle

`e-spreading -- the increase in pulse width within a given
`ber due to effects of modal and material dispersion

Basic Definitions (cont.)

Rayleigh scattering -- a scattering of the incident radiation through a fiber inversely proportional to the fourth power of the wavelength caused by various heterogeneities in the fiber

refractive index -- the ratio of the velocity of light in a vacuum to its velocity in the core or cladding of the fiber

signal-to-noise ratio -- the ratio of the power in the transmitted signal to the undesirable noise present in the absence of any signal

single mode fiber -- an optical fiber of core diameter d whose value V is <2.4 where $V = \pi d(NA)/\lambda$ where λ is the wavelength of the transmitted light

step-index fiber -- an optical fiber whose core has a uniform refractive index up to the inner edge of the cladding

telecommunication -- classical communications by electrical transmission including telephone, telegraph and television

total internal reflection -- the reflection occurring within the fiber waveguide when the angle of incidence of light striking the surface is greater than the critical angle

YAG laser -- the solid-state laser of small surface area and volume composed of yttrium aluminum garnet

Fiber Optic Test Methods

The following test methods are presented in abbreviated format from the proposed military standards for fiber optics test methods prepared by the Fiber Optics Task Group of the Society of Automotive Engineers SAE A2H, dated December 1976. (Not included is MIL STD 1553/FO being prepared by the A2K subcommittee as a general purpose MUX Specification.) During the next few years, these test methods will be presented in a more complete and finalized format after more experience is gained in testing the properties of fiber optic cables and fiber optic components. Complete procedures for testing light sources (lasers and LED's), detectors (PIN's and APD's) and other fiber optic communications components have not yet been formalized. Particularly with light sources, the user should be cautious of any reported values by the manufacturer such as light intensity and life-time-in-use since these test methods have not yet been determined on an international basis. With detectors, the reported values for responsivity, noise and life-time-in-use should also be viewed cautiously.

When reading the established test methods for fiber optic cables, the user should keep in mind that the most important values for all tests are those for the actual expected environmental conditions to which the cable will be subjected. For example, the elongation and fiber integrity properties at elevated temperatures will be much more significant for telecommunications cables strung out on poles in hot South-East Asian climates than for cables lain inside underground duct work in Canadian installations. Part of the purpose of the fiber optics standards and test methods is to enable the prospective buyer to choose the cable that best suits a particular application both in terms of environmental considerations, data rate capacity and expected repeater spacings as well as compatibility with the other fiber optics components (light sources, splices, detectors, couplers, signal splitters, etc.).

Cyclic Flexing - 2010

This test method describes a procedure for determining the ability of a fiber optics cable to withstand cyclic flexing. The specimen used for the test should be a representative sample of the finished cable product.

Apparatus

The apparatus used for this test is simply a motor drive assembly to rotate the cable 30 times per minute through a half-circle swing of about a ten inch radius. A weight is attached to the bottom the of cable specimen, typically in 1 kg increments. Clamps are used to minimize the vertical movement of the cable during the test cycle.

Procedure

1. Prior to the cyclic flexing test, the number of whole or transmitting fibers is determined. For single fiber cables this is rather simple. For multi-fiber bundles several counting methods should be used to precisely determine the number of fibers in good transmitting condition.

2. The cable sample is preconditioned at $23 \pm 2^{\circ}C$ and $50 \pm 5\%$ relative humidity.

3. The specimen is taken from the preparation chamber to the flexing apparatus and secured in position with the appropriate weights added to the lower vertical segment.

4. The flexing drive gear is then activated for a given number of cycles (usually in increments of 30).

5. The test cable is then removed from the equipment and again tested for the number of transmitting fibers. For many types of cables, it is possible to break a fiber a still have some degree of transmission through the fibers. For larger fiber bundles the actual output power should be measured before and after each of the flexing tests to more precisely determine the partial breaks in the cable. (See Method 6010).

6. This general procedure can be repeated with different weights using various numbers of flexing cycles.

Results

Test results should be reported by recording the fiber type, the weight used for each cyclic test, the number of cycles used, the percentage of broken fibers and/or the percentage change in ouput as measured by a specific detector at a specified wavelength.

Cable Twist - 2050

This test method describes a procedure for determining the ability of a fiber optics cable to withstand rotational stress by determining the number of broken fibers and/or the change in the fiber attenuation after each set of applied stresses. The sample chosen for the test should be representative of a particular cable batch. The test should be repeated several times for precision.

Apparatus

The equipment used for this test basically consists of some clamping devices to hold one part of the cable stationary while the other end is rotated 180° clockwise then 180° counter-clockwise for a given number of cycles.

Procedure

1. Prior to the twisting tests, the number of transmitting fibers should be determined and the fiber attenuation measured as given in Method 6020.
2. The cable sample is attached to the apparatus and the temperature is recorded and controlled if necessary to some specific level.
3. The torquing clamp is then rotated $\pm 180^{\circ}$ for ten cycles.
4. The fiber or fiber bundle attenuation is then measured and the number of transmitting fibers is determined.
5. The entire test sequence can be repeated several times at different temperatures if desired. In all physical tests on fiber cables the expected extreme temperatures in actual usage should be examined to establish the best and worst case results.

Results

For the cyclic cable twisting tests the results should include the details of the fiber cable specifications, the temperature used for each test, the number of cycles of applied stress, the fiber attenuation before and after each test cycle, the wavelength used to determine attenuation, the specification of the detector used, and the number of twists required to attain complete loss of the cable transmission. (Note that in some cable configurations where the fiber is contained loosely inside of internal tubes, the number of twists required to break off transmission may be very large and beyond the reasonable time limit allotted for such a test).

Power Transmission Vs. Temperature - 4010

This test method describes a means for determining the effect of temperature on the transmitted power in a fiber optic cable. The temperature effect on the transmitted power is given by : $\tau = \Phi_2/\Phi_1$, where Φ_2 is the radiant power at temperature T_2 and Φ_1 is the radiant power at T_1. T_1 is usually specified at 25°C.

Apparatus

The equipment used for this method is basically the same as is used for method 6010. The only difference is that the specimen is placed inside of a temperature controlled chamber which should not interfere with the light source or detector units. Thermocouples should be used to accurately record and monitor the temperature of the specimen and the temperature of the chamber unit.

Procedure

1. The output of the light source is adjusted to the center wavelength, bandwidth and specified intensity.
2. Both ends of a long length of the fiber cable are prepared with smooth flat end surfaces and properly aligned with the source and detector. Index matching fluid can be used if needed.
3. The test specimen is first measured at ambient conditions which should be near T_1 as specified. The actual specimen and chamber temperatures are recorded.
4. The relative (or absolute) radiant power is measured at the given wavelength. If needed, a set of wavelengths can be used to measure the transmission radiance at each temperature.
5. The temperature of the chamber is then changed to a higher temperature, and the radiant power is again measured. At each temperature change the specimen should be allowed to reach a thermal steady state, usually achieved within four hours.
6. The ratios of the transmitted power are then calculated for each set of wavelenghts and temperatures.

Results

After the completion of the measurements the test results should include the calculated ratios, the wavelengths and bandpass of the light source, the total length of the fiber cable specimen and the actual heated length within the chamber, the index matching fluid description, the type and responsivity of the detector used at the given wavelengths, and if needed, the number of transmitting fibers before and after each heat conditioning period.

Radiant Power Measurements - 6010

The purpose of this method is to describe the test to measure the total radiant flux (power) emanating from a source or from
a fiber cable. The radiant flux is given in watts (Joules/second).
When continuous measurements are used, the radiant power represents
a constant value. When synchronous or tuned amplifier techniques
are used, the reported radiance is proportional to the total
eminating radiant power.

Apparatus

The equipment used for this test method basically consists of
a radiation source, optics to control and define the eminating light
and a radiation detector and power measurement system. The light
source can be a laser, LED or lamp depending upon the desired wavelength and launch cone. The source must be stable in time and uniform in spatial intensity. Optical filters and diffraction gratings
can be used to adjust the bandpass and center frequency. An iris
diaphragm and lens system can also be used to remove stray light to
the detector and adjust the beam shape and launch pattern. The
detector can either be a large area photodetector or a calibrated
integrating sphere. When synchronous detection is used, the power
meter apparatus used may include a tuned or phase locked amplifier,
using an external light modulator as a reference beam.

Procedure

1. The output of the light source is adjusted to the desired
wavelength, bandwidth and intensity.
2. The specimen is placed in line with the detector. Some
index matching fluid can be used between source, specimen and the
detector. Fiber specimens must be prepared with perpendicular ends.
3. For continuous power measurements the relative (or
absolute) radiant power is measured at the specified wavelengths.
For tuned measurements the rms value of the component of the modulated radiant power is taken at the specified wavelengths.

Results

The various tests performed for radiant power measurements
should report the total power (in watts), whether the power is a
relative or absolute value, the radiation source, center wavelengths
used, bandpass at the wavelengths, NA of the launched radiation cone
and the optical system used, the length of the specimen, a description of the index matching fluids and the detector system used.

Attenuation Measurement - 6020

This test outlines the procedure for measuring the attenuation of a specified length of fiber cable at specified wavelengths, by taking radiant power measurements at two known fiber lengths. The attenuation in dB/km is defined by : $\beta = -10 \log(\Phi_1/\Phi_2)/\Delta L$, where Φ_1 is the radiant power at length L_1, Φ_2 is the radiant power at length L_2 and $\Delta L = L_1 - L_2$.

Apparatus

The equipment used for this procedure is essentially the same as is used for the previous method (6010). For the attenuation measurements it is important that a uniform Lambertian source is used to excite all fiber modes equally. In both the radiant power and in the attenuation measurements it is important to keep applied stress from the optical cable which may induce bending or stress losses in the fibers. For large lengths of cable the holding reel should be at least two feet in diameter and preferably greater for cables with an outside diameter above 2 cm.

Procedure

1. After the number of transmitting fibers is determined, the output of the light source is adjusted to the center wavelength and intensity and bandwidth specified.
2. Both ends of the cable are carefully prepared with flat perpendicular end surfaces, with index matching fluid if needed, between source - specimen - detector.
3. The relative (or absolute) radiant power is measured at a particular wavelength or set of wavelengths for both fiber lengths.
4. The procedure can be repeated with another set of cable specimens from the same batch to confirm the attenuation measurement and calculation, which should be close to the manufacturer claims.

Results

The test results for this procedure should include the number of transmitting fibers in the cable, the center wavelength, bandpass and numerical aperture of the radiation source and launch cone, the lengths of the cables for each measurement, the radiant power at each length and wavelength, the index of refraction of the fluids used, the type, size and responsivity of the detector used at each wavelength, and the calculated attenuation for each set of measurements. If used, the type of cladding mode stripper should be noted.

Radiation Pattern Measurement - 6030

The purpose of this procedure is to measure the radiation
pattern of a fiber cable by determining the transmitted power as a
function of the output radiation launch angle. This same procedure
could be used to determine the radiation pattern of light sources.
For either measurement, either the source or detector is fixed
in position and the other rotated in a single plane. For known
asymmetrical specimens the test should be expanded to some form of
a three-dimensional radiation pattern measurement.

Apparatus

The equipemnt used for this method is essentially the same
as is used for method 6010. The major difference is that the angle
between source and detector is measured using a goniometer and X-Y
recorder to plot angle versus radiant power. The movement of the
goniometer should permit accurate measurements over $\pm 90^\circ$. For
these measurements either the detector or specimen should remain
fixed. The following procedure is for a moveable radiation detector which is pivoted about the source (fiber end or source).

Procedure

1. The output of the source is adjusted to the center wavelength for fiber cables or simply activated for light sources, and
the intensity and bandwidth adjusted to the desired levels.
2. The finished end of the cable or the surface of the
light source specimen is fixed to the goniometer base, to establish
the zero position for the recorder measurements.
3. The detector is fixed to the moveable end of the goniometer arm and positioned such that with the arm at the arbitrary
zero position it is coincident with the axis of the specimen.
4. The relative (or absolute) radiant power is measured as a
function of angle which is accomplished quickly using an X-Y recorder which can be previously calibrated for absolute values.

Results

The results of the radiation pattern plots should include
the type of light source, the center wavelength and bandwidth, the
type of specimen used and its preparation, the numerical aperture
of the launching radiation cone, the radiation plot of intensity
vs. angle and whether the detector or specimen was held fixed. For
three-dimensional measurements all angles must be specfied to allow
the complete spatial distribution to be plotted.

Pulse Spreading - 6050

This test method describes the procedure for measuring the pulse spreading in a fiber optics cable defined as :
$\Delta(50) = (W_1^2 - W_2^2)^{1/2} / (L_1 - L_2)$, where W_1 is the pulse width at 50% maximum pulse amplitude of the test specimen output wave form, L_1 is the length of the test specimen and W_2 and L_2 refer to the reference specimen, which is a short length of the test specimen.

Apparatus

The equipment used for this test is about the same as in 6010 except that the light beam is split into two parts using two light detectors. In addition, a pulse recorder such as a dual beam storage oscilloscope, is used to display pulse amplitude vs. time at the detectors. A camera or X-Y recorder can be used with the scope to attain a permanent record of the output waveforms of the modulated light source. If a beam splitter is not used the test may be done by substituting the specimens under identical conditions.

Procedure

1. The output of the source is adjusted to the specified center wavelength, bandwidth and intensity and coupled to the cable.
2. The other end of the fiber specimen is coupled to the detectors such that the flat end is perpendicular to the axis of the impinging light.
3. The maximum amplitude of the light pulses are adjusted to be approximately equal, by adjusting the beam splitter ratio or by using an attenuator in the reference beam.
4. The output wave forms of both specimens (both beams) is recorded for permanent record. The pulse widths (in nanoseconds) are measured at 50% maximum pulse amplitude. The source pulse should be adjusted to make $W_1 \leq 1.4\ W_2$.

Results

After completion of the tests the results should include the type of light source, the center wavelength and bandpass, the length of the test and reference specimens, the type of index matching fluid used, the type and responsivity of the detectors, the pulse widths at 50% maximum amplitude, the calculated pulse spreading, the output waveform and the method of spooling the fiber cable. If several wavelengths are used during the tests each set of pulse widths should be reported with a permanent record of the individual pulse shapes for each beam.

Fiber Transfer Function - 6070

The purpose of this test procedure is to measure the transfer function of a fiber or fiber bundle. The transfer function is given by : FTF = H(f) = A(f) exp(iθ(f)), where A(f) is the amplitude transfer function (AFT) and θ(f) is the phase transfer function, (PTF). Both a test and reference specimen of cable are used, where the reference sample is a short length of the test specimen.

Apparatus

The basic experimental set-up used for this test is similar to that used for method 6010. A sweep frequency generator is used to modulate the source and a beam splitter is used to divide the source light into two beams, detected by two identical detectors. A calibrated recorder such as a vector volt-meter is used to measure the detector ouputs with respect to both phase and amplitude. A suitable X-Y recorder is used to record the output voltage vs. frequency and the phase angle vs. frequency.

Procedure

1. The output of the light source is adjusted to the center wavelength, bandwidth and specified intensity.
2. The fiber specimens must be carefully prepared with the end faces smooth and flat and perpendicular to the fiber axis.
3. The two fiber samples are aligned in each beam with ends coupled to the light sources and detectors. Index matching fluid can be used to couple the specimens to the radiation source and to the detectors.
4. The voltage of the detectors is adjusted such that they are in the linear range of the detectors by varying the beam ratios.
5. For each channel the output amplitude $A_1(f)$, $A_2(f)$ and phase $\theta_1(f)$ and $\theta_2(f)$ are recorded as a function of frequency f.

Results

After completion of this test the report on the results should include the type of light source, the center wavelength and bandpass, the NA of the launched radiation cone, the lengths of the fiber specimens, the type of index matching fluid used, the type, and responsivity of the detectors at the specified wavelengths, the computed fiber transfer functions and the test curves of voltage and phase vs. frequency.

Refractive Index Profile - 6090

This procedure describes the method to measure the refractive index profile of a graded or step index fiber using a near field test. Using long lengths of the fiber or fiber cable, such that the length is greater than 2 meters, the refractive index is given by : $\Delta n^2(r) = (NA)^2 I(r)/I(o)$, where n is the change in refractive index, NA is the numerical aperture, $I(r)$ is the intensity of the radiation measured at a distance r from the pattern center and $I(o)$ is the intensity maximum at the pattern center.

Apparatus

The equipment used for this method is similar to that used in method 6010. It is important in this test that the light source be Lambertian and that the lauch system be capable of filling the entire numerical aperture of the fiber specimen. The ouput radiation is focused onto a plane for observation. An X-Y recorder may be coupled to the detector to plot intensity vs. detector position. The NA of the detector should be no smaller than the NA of the output lauch system.

Procedure

1. The NA of the fiber is first determined using test method 6030.
2. The detector is then positioned such that it lies in the focal plane of the specimen image throughout the scanning distance.
3. The radiant power is measured using method 6010 while the detector is transversed across the major diameter of the specimen image, and recorded as a function of position.
4. The specimen is then rotated and the radiant power measured as the detector transverses the minor diameter and recorded as a function of the detector position.

Results

For this test procedure the results should include the change in refractive index for both the major and minor diameter scans, the type, size and NA of the detector, the length, NA and major and minor diameter of the specimen, and the wavelength and bandpass of the light used as the Lambertian source. Most fibers should exhibit symmetrical profiles as a function of radius (the distance from the fiber center). When different materials are used in the core and cladding absorption changes can be wavelength dependent, requiring that several wavelengths be used for the test.

Company Addresses and Products*

Aborn Electronics
1928 Old Middlefield Road
Mountain View, Calif. 94043
(connectors,detectors,LED's)

AEG-Telefunken
P.O. Box 1109
D-71 Heilbronn, West Germany
(cables,detectors,LED's,systems)

American Laser Systems
106 James Fowler Road
Goleta, Calif. 93017
(detectors,lasers,modules,
 systems)

AMP, Inc.
Harrisburg, Penn. 17105
(connectors)

Amphenol Optical Products
33 East Franklin Street
Danbury, Conn. 06810
(connectors,crimpers,splices)

Anaconda Wire And Cable
Greenwich Office Park
Greenwich, Conn. 06830
(cables)

Asahi Chemical Industry Co.
HibiyaMitsui Building
1-2 Yuraku-cho, 1-Chome
Chiyoda-ku Tokyo 100, Japan
(plastic fibers and cables)

ASEA-HAFO
Allmanna Svenska Elektiska AB
Vallingby, Sweden
(detectors,LED'slinks)

Bell Canada
620 Belmont Street
Montreal, Quebec, Canada
(cables,connectors,detectors,
 LED's,links,modules,splices)

Belling & Lee, Ltd.
C & I Systems Division
Great Cambridge Road
Enfield, Middlesex, England
(cables,connectors,links)

Bell Telephone Laboratories
Mountain Avenue
Murray Hill, N.J. 07974
(telecommunication systems)

Berg Electronics
Division Of Dupont
New Cumberland, Penn. 17070
(connectors)

Boeing Company
Seattle, Washington 98124
(aircraft systems)

Breeze-Illinois, Inc.
Main & Agard Streets
Wyoming, Illinois 61491
(connectors)

British Callender Cables, Ltd.
P.O. Box 9
Prescot, Lancashire, U.K.
(cables)

Cables De Lyon
170 Avenue Jean Jaures
F69007 Lyon, France
(cables,connectors)

Canada Wire & Cable, Ltd.
147 Laird Drive
Toronto 352, Ontario, Canada
(cables)

*For complete product listings, consult manufacturers.

Company Addresses and Products (cont.)

Canon, Inc.
9-9 Ginza, 5-chome Chuo-ku
Tokyo, Japan
(cables)

Centronic
1101 Bristol Road
Mountainside, N.J. 07092
(cables,detectors,links,modules)

Clairex Electronics
560 S. Third Avenue
Mt. Vernon, N.Y. 10550
(detectors)

Collins Radio Company
1200 North Alma Road
Richardson, Texas 75080
(systems)

Corning Glass Company
Houghton Park CO7
Corning, N.Y. 14830
(cables,connectors,links)

Dainichi Nippon Cables, Ltd.
Umeda Building, 7-3 Umeda
Kita-ku, Osaka, Japan
(cables)

Decca, Ltd.
Ingate Place
Queenstown Road
London SW8, England
(systems)

Devar, Inc.
Control Products Division
706 Bostwick Avenue
Bridgeport, Conn. 06605
(detectors,modules)

Digital Equipment Corp.
146 Main Street
Maynard, Mass. 01754
(computer systems)

Dolan-Jenner Industries, Inc.
200 Ingalls Center
Melrose, Mass. 02176
(cables,detectors,links)

Dupont De Nemours & Company
Plastic Products & Resins
Wilmington, Delaware 19898
(plastic fibers and cables)

The Ealing Corporation
22 Pleasant Street
South Natck, Mass. 01760
(detectors,fiber bundles)

EG & G, Inc.
Electro-Optics Division
35 Congress Street
Salem, Mass. 01970
(detectors,modules)

Electro-Fiber-Optics Company
99 Hartwell Street
West Boylston, Mass. 01583
(cables,connectors)

ElectroOptic Components & Controls
15 Tech Circle
Natick, Mass. 01760
(cables,detectors,lenses,systems)

Electrophysics Corporation
48 Spruce Street
Nutley, N.J. 07110
(detectors,systems)

EMI Electronics, Ltd.
Hayes, Middlesex, England
(detectors)

Ericsson, L.M.
Vag 4-8 Fack 12611
Stockholm 32, Sweden
(telecommunications systems)

Company Addresses and Products (cont.)

Fairchild Semiconductor
Optoelectronics Division
4001 Miranda Avenue
Palo Alto, Calif. 94304
(LED's)

Ferranti, Ltd.
Dunsinane Avenue
Dundee DD23PN, U.K.
(systems)

Fiber Optic Cable Corp.
Framingham, Mass. 01701
(cables)

Fort
15 Rue d'Argenteiul
75001 Paris, France
(cables,detectors,links)

Fujikura Cable Works, Ltd.
Kasumigaseki Building
2-5 Kasumigaseki, 3-chome
Chiyoda-ku, Tokyo, Japan
(cables)

Fujitsu, Ltd.
6-1 Marunouchi, 2-chome
Chiyoda-ku, Tokyo, Japan
(connectors,detectors,lasers,
 LED's,links,modules,repeaters)

Furukawa Electric Company
6-1 Marunouchi, 2-chome
Chiyoda-ku, Tokyo 100, Japan
(cables,connectors,links,splices)

FYBTEK
3549 Haven Avenue
Menlo Park, Calif. 94025
(cables)

Galileo Electro-Optics Corp.
Galileo Park
Sturbridge, Mass. 01518
(cables,connectors,detectors,
 LED's,links,modules)

General Cable Corporation
500 West Putnam Avenue
Greenwich, Conn. 06830
(cables)

General Electric Company
570 Lexington Avenue
N.Y., N.Y. 10022
(detectors,industrial systems)

General Instrument
1775 Broadway
N.Y., N.Y. 10019
(detectors)

General Optimation, Inc.
1340 Post Road
Southport, Conn. 06490
(systems)

✗ GTE Labs, Inc.
40 Sylvan Road
Waltham, Mass. 02154
(telecommunications systems)

Harris Corporation
55 Public Square
Cleveland, Ohio 44113
(data buses,links,modulators,
 systems)

Hawker-Siddeley Aviation, Ltd.
Kingston Road
Kingston-Upon Thames
Surrey, U.K.
(aircraft systems)

Hellerman-Deutsch
East Grinstead
Sussex RH191RW England
(cables,connectors,links,splices)

Heraeus-Schott Quarzschmelse
6450 Hanau
West Germany
(cables)

Company Addresses and Products (cont.)

Hewlett Packard
Optoelectronics Division
640 Page Mill Road
Palo Alto, Calif. 94304
(detectors,lasers,LED'ssystems)

Hitachi Cable, Ltd.
New Marunouchi Building
1-5 Marunouchi, 1-chome
Chiyoda-ku, Tokyo, Japan
(cables,lasers,LED's)

Hoya Glass Works, Ltd.
Shinjuku Dai-ichi Seimei Bldg.
24-1 Nichi-Shinjuku, 1-chome
Shinjuku-ku, Tokyo, Japan
(cables)

Hughes Aircraft Company
Hughes Research Laboratories
Centinela Avenue & Teale Sts.
Culver City, Calif. 90032
(aircraft systems,detectors)

IBM
Manchester Bridge
Poughkeepsie, N.Y. 12603
(computer systems)

Infrared Industries, Inc.
62 Fourth Avenue
Waltham, Mass. 02154
(detectors)

International Light, Inc.
Dexter Industrial Green
Newburyport, Mass. 01950
(test equipment)

IRT Corporation
Fiber Optics Division
7650 Convoy Court
San Diego, Calif 92111
(plastic fibers and cables)

Isomet
5414 Port Royal Road
Springfield, Virginia 22151
(modulators,switches)

ITT
320 Park Avenue
N.Y., N.Y. 10022
(cables,connectors,detectors,
lasers,LED's,links,repeaters,
test equipment)

ITT-Canon Electric
666 East Dyer Road
Santa Ana, Calif. 92705
(connectors,couplers,systems)

Jenaer Glaswerk Schott & Gen.
P.O. Box 2480
6500 Mainz, West Germany
(cables,connectors)

Keystone Optical Fibers
120 Guild Street
Norwood, Mass. 02062
(cables)

Kollmorgen Corporation
60 Washington Street
Hartford, Conn. 06106
(submarine systems)

Laser Diode Laboratories
205 Forrest Street
Metuchen, N.J. 08840
(lasers,LED'slinks)

Litronix, Inc.
19000 Homestead Road
Cupertino, Calif. 95014
(LED'smodules)

Lockheed Palo Alto Res. Labs
3251 Hanover Street
Palo Alto, Calif. 94304
(satellite systems)

Company Addresses and Products (cont.)

Lomax Company
671 Enchanted Way
Pacific Palisades, Calif. 90272
(fiber bundles)

Marconi-Elliot Avionic Systems
Elstree Way
Borehamwood, Herts, U.K.
(aircraft systems)

Material Telephonique
46-47 Quai Alphonse Le Gallo
Boulogne-Billancourt
Haut-Sen 92100, France
(telecommunications systems)

Matsushita Communication Co.
1006 Kadoma
Osaka, Japan
(cable television systems)

McDonnell Douglas Corporation
P.O. Box 516
St. Louis, Mo. 63124
(satellite systems)

Meret, Inc.
1815 24th Street
Santa Monica, Calif. 90404
(connectors,detectors,LED's,
 links,modules,systems)

Mitsubishi Electric Corp.
2-3 Marinouchi, 2-chome
Chiyoda-ku, Tokyo 100, Japan
(cables,plastic fibers,systems)

Monsanto Commercial Products
Electronics Division
3400 Hillview Avenue
Palo Alto, Calif. 94304
(LED's)

3M Company, 3M Center
St. Paul, Minn. 55101
(connectors,plastic fibers)

National Semiconductor
2900 Semiconductor Drive
Santa Clara, Calif. 95051
(LED's)

Newport Research Corp.
18235 Mt. Baldy Circle
Fountain Valley, Calif. 92708
(manipulators)

Nippon Electric Company
7-15 Shiba, 5-chome
Minato-ku, Tokyo 108, Japan
(cables,connectors,detectors,
 lasers,LED's,power systems,
 telecommunications systems)

Nippon Sheet Glass Company
4-Chome, Dosho-machi
Higashi-ku, Osaka, Japan
(cables)

Nortronics
Division Of Northrop
1 Research Park
Palos Verdes Pen., Calif 90274
(aircraft systems)

Optelecom, Inc.
15940 Shady Grove Road
Gaithersburg, Maryland 20760
(cables,links,systems)

Optical Communications Corp.
950 Norwood Road
Silver Spring, Maryland 20904
(links)

Opticom
3600 M Street N.W., Suite 400
Washington, D.C. 20007
(systems)

Optics For Research
Box 82
Caldwell, N.J. 07006
(cables,connectors)

Company Addresses and Products (cont.)

Optoelectronics, Inc.
1309 Dynamic Street
Petahima, Calif. 94952
(detectors)

Opto Micron Industry Company
Sagami Building 7-11, 5-chome
Yaesu, Chuo-ku
Tokyo 104, Japan
(connectors,manipulators,
polishers,splices)

Optron, Inc.
1201 Tappan Circle
Carrollton, Texas 75006
(detectors,LED's,systems)

Phillips
Gloeilampenfabrieken
Enhoven, The Netherlands
(detectors,lasers,LED's)

Pilkington Brothers, Ltd.
Prescot Road
St. Helens, Lancashire, England
(cables)

Plessey Telecommunications
Edge Lane
Liverpool L79NW, England
(detectors,lasers,LED's,links)

Poly-Optics, Inc.
1815 E. Carnegie Avenue
Santa Ana, Calif. 92705
(cables,plastic fibers)

Quadri Corporation
2950 West Fairmont
Phoenix, Arizona 85017
(cables,computer systems,
 detectors,links)

Quantrad Corporation
2261 S. Carmelina Avenue
Los Angeles, Calif. 90064
(detectors,modules)

Quartz & Silice
8 Rue D'Anjou
Paris 75008, France
(cables)

Quartz Products Corporation
608 Somerset Street
Plainfield, N.J. 07061
(cables)

Radiation Devices Company
P.O. Box 8450
Baltimore, Maryland 21234
(cables,connectors,LED's
 modules,systems)

Rank Precision Industries, Inc.
P.O. Box 332
260 North Route 303
West Nyack, N.Y. 10994
(cables)

Raytheon Company
141 Spring Street
Lexington, Mass. 02173
(detectors,systems)

RCA - Solid State Division
Electro Optics And Devices
Lancaster, Penn. 17604
(detectors,lasers,LED's,links)

Rockwell International
3430 E. Mirloma Avenue
Anaheim, Calif. 92806
(systems)

Sealectro Corporation
Mamaroneck, N.Y. 10543
(connectors)

Showa Electric Wire & Cable
10 Shiba Toranomon
Minto-ku, Tokyo, Japan
(cables)

Company Addresses and Products (cont.)

Siecor
Kistlerhofstrasse 170, D-8000
Munchen 70, West Germany
(cables,connectors,links)

Siemens-Aktiengesellschaft
Wittelsbacherplatz 2 D-800
Munchen, West Germany
(telecommunications systems)

Smiths Industries, Aviation Div.
Edgeware Road, Cricklewood
London NW26JN, U.K.
(aircraft systems)

Spectronics, Inc.
830 E. Arapaho Drive
Richardson, Texas 75081
(cables,connectors,detectors,
 lasers,LED's,links,systems)

Sumitomo Electric Industries
The New Sumitomo Building
15, 5-Chome Kitahama
Higashi-ku, Osaka, Japan
(cables,systems)

Tektronix, Inc.
P.O. Box 500
Beaverton, Oregon 97077
(LED's, test equipment)

Telecommunications R.E.T.
26 Rue Boyer
Paris 2E, France
(systems)

Telephone Cables, Ltd.
190 Strand
London, WO2RIDU, England
(systems)

Texas Instruments, Inc.
P.O. Box 5474
Dallas, Texas 75222
(detectors,LED's)

Thomas & Betts Corporation
3208 Humboldt Street
Los Angeles, Calif. 90031
(connectors)

Thomson & CSF
173 Boulevard Haussman
Paris, France
(cables,connectors,detectors,
 LED's,links)

Times Fiber Communications
358 Hall Avenue
Wallingford, Conn. 06942
(cables,lasers,LED's,links)

TRW/Cinch Connectors
1501 Morse Avenue
Elk Grove, Illinois 60007
(connectors)

Twentieth Century Electronics
King Henry's Drive
New Addington, Croydon, U.K.
(connectors,detectors,links)

United Aircraft Research Labs
400 Main Street
E. Hartford, Conn. 06108
(aircraft systems)

United Detector Technology
2644 30th Street
Santa Monica, Calif. 90405
(detectors,modules)

Valtec Corporation
99 Hartwell Street
West Boylston, Mass. 01583
(cables, connectors)

Vought Systems Division
9314 W. Jefferson
Dallas, Texas
(aircraft systems)

INDEX